THE HUMAN ODYSSEY

In this panoramic scene artist Jay Matternes has re-created a representative primate from each of four points during the evolution of this group. The environment of each also has been meticulously reconstructed. From right to left, the primates are: *Notharctus* (48 million years old), *Aegyptopithecus* (34 million years old), *Proconsul* (18 million years old), and *Sivapithecus* (8 million years old).

THE HUMAN ODYSSEY

FOUR MILLION YEARS OF HUMAN EVOLUTION

IAN TATTERSALL, CURATOR
AMERICAN MUSEUM OF NATURAL HISTORY

FOREWORD BY DONALD C. JOHANSON

Prentice Hall General Reference
New York London Toronto Sydney Tokyo Singapore

A New England Publishing Associates Book
Designer: Don Brunelle
Copy Editor, Indexer: Larry Hand
Administrator: Susan Brainard

Prentice Hall General Reference
15 Columbus Circle
New York, NY 10023

Library of Congress Cataloging-in-Publication data
Tattersall, Ian.
The human odyssey: four million years of human evolution/Ian Tattersall;
with a foreword by Donald C. Johanson.
p. cm.
"Based on the acclaimed *Hall of Human Biology and Evolution*
at the American Museum of Natural History."
Includes bibliographical references and index
ISBN 0-671-85005-9
1. Human evolution.   I. American Museum of Natural History. II. Title.
GN281.T36 1993
573.2—dc20   92-21341   CIP

Manufactured in the United States of America

10 9 8 7 6 5 4 3 2 1

First Edition

This volume is dedicated to
Willard Whitson,
exhibition designer extraordinary,
and to all those many other gifted individuals
who contributed to the
Hall of Human Biology and Evolution
at the
American Museum of Natural History.

Acknowledgments

This book grew out of the American Museum of Natural History's new *Hall of Human Biology and Evolution*, whose aim is to show where our own human species fits into the natural world. Both book and exhibition are thus concerned with the story of our evolution from the very beginning of life on Earth, and with how we understand that evolution. This is a dramatic story indeed, and while this book is able to draw on visual elements prepared for the exhibition that evoke that drama in a unique and powerful way, it also allows fuller discussion of many themes that could only be touched on briefly in the exhibition itself.

Any exhibition is a corporate effort, and as curator of the *Hall of Human Biology and Evolution* I owe a large debt to many people. This book gives me an opportunity at least to acknowledge this debt, if not to repay it. Without the contributions of the following academic friends and colleagues (listed in alphabetical order), the *Hall* could never have approached the excellence it has achieved: Jelle Atema, Bob Brain, Dodi Ben-Ami, Rob Blumenschine, Jaymie Brauer, Tim Bromage, Dominique Buisson, Ron Clarke, Jean-Jacques Cleyet-Merle, Hilary Deacon, Eric Delson, Andrea Dunaif, Niles Eldredge, Craig Feibel, Phil Gingerich, Fred Grine, Gregg Gunnell, John Holmes, Sid Horenstein, the late Alun Hughes, Jan Jelinek, Peter Jones, James Kitching, Jeff Laitman, Henry and Marie-Antoinette de Lumley, Alex Marshack, Jay Matternes, André Morala, Hansjürgen Müller-Beck, David Pilbeam, Yoel Rak, Jean-Philippe Rigaud, Mike Rose, Alain Roussot, Jeff Schwartz, Olga Soffer, Sam Taylor, Francis Thackeray, Alan Thorne, Phillip Tobias, Jean-François Tournepiche, John Van Couvering, Alan Walker, Ward Wheeler, and Randy White. In this category I should also include Willard Whitson, who went far beyond his role of designer of the *Hall* to affect many aspects of its content as well, and to whom are due many of the photographs in this book. And of course I should not forget all of those many other talented members of the exhibition, scientific, construction, and administrative departments of the American Museum of Natural History who worked with dedication and inspiration to make this great exhibition possible. You are too numerous to thank here individually, but you know who you are. Thank you all.

The successful realization of the *Hall* owes much to the generosity of many individuals and organizations. Among these I would most particularly like to acknowledge the following: the Bristol-Myers Squibb Company, the National Science Foundation (Informal Science Education Program), the Richard Lounsbery Foundation, the Booth Ferris Foundation, the New York Times Company Foundation, the Charles Hayden Foundation, and the Lila Acheson Wallace Fund/New York Community Trust.

My appreciation goes also to Donald C. Johanson, who kindly contributed the Foreword to this volume. Don is one of the most inspired fossil-finders in the annals of paleoanthropology, and his discoveries have transformed our understanding of the earliest evolution of our kind.

The illustrations in this book were furnished by many kind colleagues and friends. All are individually acknowledged in the captions, but here I would like to express my most especial gratitude to Diana Salles, Jay Matternes, Willard Whitson, Don McGranaghan, and Chris Rossi.

And finally, a word of profound thanks and appreciation to Deirdre Mullane, my editor at Prentice Hall, to Elizabeth Frost Knappman and Ed Knappman, of New England Publishing Associates, to Larry Hand, and to designer Don Brunelle, without whom this book would never have seen the light of day.

Ian Tattersall
American Museum of Natural History

Contents

Foreword

The concept of evolution stands as one of the greatest ideas ever formulated by the human mind. Although the notion of evolution flickered through the minds of many, it was Charles Darwin who, in 1859 in his *On the Origin of Species*, provided the first cogent argument for "descent with modification." When Darwin left England in December of 1831, as the ship's naturalist on H.M.S. *Beagle* on a round-the-world voyage, he was only twenty-two years old. He had little idea that the voyage would be a revelation to him and would result in a profound revision of his view of the natural world. We must not forget that, in the early nineteenth century, the accepted belief was that all life was divinely created in its present form, and immutable. Darwin himself subscribed to this belief when he first stepped aboard the *Beagle*. But by the time he had completed his voyage five years later, he knew he had been wrong.

Darwin was uniquely equipped for his voyage. He was uncommonly observant; he had a strong natural history background in zoology, botany, and geology; and, most importantly, he did not have any particular scientific axe to grind. His mind was open, and he noted everything he saw. He was a man of remarkable insight, and after his great journey around the world, he pondered long over what he had observed. He recognized a thread that bound all life; and he eventually concluded that all life, extant or extinct, had been shaped by the forces of natural selection. But the most revolutionary implication of his theory—which provoked great controversy among his contemporaries—was that humans had also been shaped by natural selection. They were no longer at the center of the natural world; rather, they were just another species in the enormous web of life.

It is fitting that the American Museum of Natural History, probably the most important museum of its kind in the world, should celebrate Darwin's ideas in its extraordinary *Hall of Human Biology and Evolution*. Through this book and the exhibit on which it is based, you will learn to better understand some of the questions all of us have asked at some time in our life: Who are we? How did we become what we are? How do we fit into the natural world? I would encourage you to study this excellent book, to marvel at the remarkable exhibits, and to learn of the molecular basis for evolution and the brilliant diversity of past and present life that the evolutionary process has produced. But I want also to give you my personal view of how vital it is that we understand our place in nature.

My interest in natural history arose when I was a young boy growing up in Connecticut and made my first visit to the American Museum of Natural History. What I saw in the halls of the Museum inspired me to look more closely at the natural world around me. I became a young naturalist—collecting butterflies, watching plants blossom and puppies being born, observing protozoa through a simple microscope. I was overwhelmed by the mystery and beauty of nature.

In high school, I was strongly influenced by an anthropologist, who frequently traveled to Africa to study the cultural practices of people living in remote areas. The more I learned about Africa, the more it fascinated me. Then, in 1959, a major human fossil discovery was announced from Olduvai Gorge in Tanzania, by the late Dr. Louis S. B. Leakey. I knew in a flash that I wanted to combine my interests in natural history and Africa, and I became a fossil hunter. I, too, would find the fossils of our ancestors!

I have had the wonderful privilege of working in Africa since 1970, first as a graduate student, then directing a number of expeditions. In 1974, while co-leading one of my very first expeditions to Africa's great Rift Valley, I had the great good fortune to find—at Hadar, in Ethiopia—a 3.2-million-year-old partial skeleton, which has since become affectionately known as "Lucy." Lucy opened a major new chapter in our understanding of human origins. There is no question that since I discovered her she has had a lasting influence on my scientific and personal life. People may not recognize my name, but mention Lucy's and people immediately think of a distant relative—which of course she is.

During subsequent expeditions to Ethiopia more and more human ancestor fossils were discovered. The site of Hadar, located in the bleak deserts of the Afar triangle, continues to yield important clues about the earliest upright-walking human ancestors. These primitive creatures, with the tongue-twisting name of *Australopithecus afarensis*, have given added weight to Darwin's prediction that Africa would ultimately prove to be the home of all mankind.

In the stunning exhibit at the American Museum of Natural History, you can see for the first time life-size reconstructions of a male and female *A. afarensis*, plodding along upright 3.5 million years ago at Laetoli, in Tanzania. This dramatic and powerful recreation will give you a remarkable glimpse of a time when our ancestors were not significantly different from any other living creature. For *A. afarensis* was truly primitive. It possessed a brain one-fourth the size of our own and, as the reconstructed models so clearly show, its face projected forward like that of a chimpanzee. But you will also notice that this creature possessed one of the most human of all characteristics—the ability to walk upright on two legs. We know this not only from the structure of the skeleton, but also because of footprints left in volcanic ash 3.5 million years ago! This interesting combination of primitive and advanced features represents a provocative link between human and ape. One can truly say that Lucy and her kind were apes who stood upright.

As you leaf through the pages of this book, you will encounter another ancestor—a more familiar one. On the wall of a cave in France an early member of our species painted an animal scene in haunting colors. Walking in the cave nearly 15,000 years ago, this person also left footprints in the mud floor—identical to our own.

These two sets of footprints span the history of our species, at least as we know it so far. Both of the makers were our precursors. One, *A. afarensis*, perhaps resembled a missing link; the other, truly human, *Homo sapiens*, was just on the brink of civilization.

The details of the fossils which fall between these two creatures are provided in this excellent book, and they are stunningly displayed in the exhibit. I will leave it to you to explore them on your own. But I would like to leave you with a few of my thoughts on the importance of knowing something about our own evolutionary origins. I have no doubt that you will lay down this book with a greatly enriched understanding of your place in nature. If we can grasp that we, too, are the consequence of natural selection, we will better comprehend our role on planet Earth. Because of the vagaries of evolutionary and climatic change, we have inherited an unprecedented responsibility to protect and preserve the very world which created us.

Evolution does not make predictions, species don't know where they're going, humans did not have to evolve. In fact, if we were to rewind the tape to ten million years ago, when apes dominated the primate world, there would be no assurance that humans would evolve again. But humans have evolved, we are here today. Like no other species that has ever lived, we control the life of all living things—including ourselves. When we understand and accept that we are part of the continuum of life, we will be in a better position to make informed choices—choices which will ensure a better world for all species. Extinction is forever. We must not let it happen.

Education is the great liberator. It frees us to think objectively. My studies of human evolution have taught me to respect the natural world. They have also taught me that all humans have a common origin and, therefore, a common destiny—the outcome of which will be determined by humankind itself. We do have the capacity to make the future a long and fruitful one, if only we will take the time to learn who we are and how we fit into the natural world.

Donald C. Johanson, Ph.D.
Institute of Human Origins

Chapter One
In The Beginning

Who are we? A deceptively simple question, though one of profound importance to a species which simply doesn't seem capable of taking itself for granted. In all of the marvelously varied human cultures around the globe, there's not one that lacks stories to account for how people came to be as they are. Perhaps this universal need stems from the fact that while we are very obviously part of the natural world, we are also set apart from it. Actually, every living species has something that distinguishes it from every other; but it is we alone who feel compelled to tell ourselves stories about why we are different.

Science is really just one more way of inventing such stories, but it differs from other kinds of story-telling in expressing them in ways that allow them to be tested—and, if necessary, rejected—by repeated experience. Scientific knowledge is thus by its nature provisional, but it provides the most reliable way we have of describing nature and how it works. The *Hall of Human Biology and Evolution* at the American Museum of Natural History seeks to satisfy our urge to understand our place in nature by telling the scientific story of who we are; and this book in turn expands on and develops the details of that fascinating story.

Our odyssey begins more than 3.5 billion years ago, with the origin of life on Earth. Our characters are the Earth's life forms, which now number anywhere from 10 million to 80 million species. (Scientists have difficulty making up their minds how many species they have not yet described.) From the beginning, living things have shown a remarkable tendency to diver-

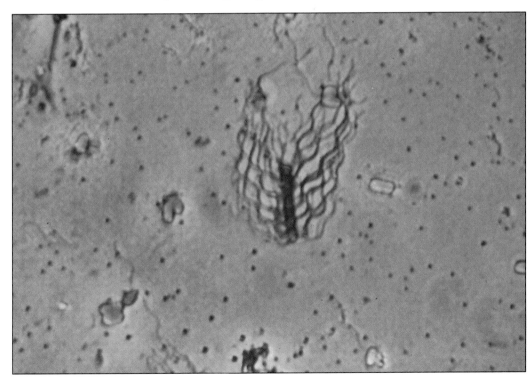

Photo taken through a light microscope of a bacterium, *Proteus vulgaris*. Such bacteria live mainly within animals but are also found in soil, water, and sewage. In humans *Proteus vulgaris* is recognized as a pathogen, associated with chronic infections of various kinds.

Micrograph by Eric Gravé; courtesy of Sam Taylor.

sify, rather as a bush branches ever outwards, with a single trunk eventually giving rise by a succession of forkings to vast numbers of twigs. In this way, every living species belongs to a great hierarchy of life that has branched out from a single ancient ancestor; and yet each has some characteristic that makes it unique.

Our place in this great tree of life is revealed in the structure of our bodies. It is here that we find the evidence for our connection to an ever-widening pool of other living things through a set of increasingly remote common ancestors. We share some of our physical characteristics with all living organisms. Others we share only with animals, or with vertebrates, or just with primates or with other hominids. Some of our characteristics are unique to ourselves, or to us and our most immediate fossil relatives. In this book, as in the *Hall of Human Biology and Evolution*, we discover our own place in nature by looking at those things we share with other living beings, and at those attributes that are unique to us. We start with the most general characteristics that all life forms have, and end with the story of how we acquired those that are singularly human.

The heritage of life

Three and a half million years ago, two early human relatives walked across a muddy plain at what is now Laetoli, in Tanzania. Discovered by paleontologists in 1976, the footprints they left behind them testify eloquently that these remote forebears were fully upright walkers. Most paleontologists believe, indeed, that our evolutionary history as human beings goes back to some point before 5 million years ago, and most of this volume is devoted to that relatively short span of geological time. Nonetheless, our history as living organisms goes back very much farther than that. For although we naturally define ourselves by those features that distinguish us from the millions of other species whose world we share, these differences are actually vastly outweighed by our similarities to other living creatures. Understanding our place in nature thus involves appreciating our shared biological heritage at least as much as our unique traits and how we came to acquire them. That's why this volume, like the exhibition on which it is based, starts right at the beginning.

All existing life is descended from a single-celled ancestor. So to discover what we human beings share most specifically with every other living thing, we must look inside the cells from which the tissues of our bodies are built. What we find is that, from the simplest bacteria on, all living organisms possess deoxyribonucleic acid (DNA). This remarkable substance is present in virtually every one of the billions of cells that make up our bodies, and it is also present in the cells of every other living organism. It is the shared possession of DNA that is the clearest proof of the common descent of all life forms on Earth.

Life's molecule

The structure of DNA—the famous "double helix"—has attained the status of folklore, and its discovery by the American biologist James Watson (1928–) and the Englishman Francis Crick (1916–) is probably the best-known biological achievement of the present century. The spiral framework of this molecule, consisting of two twining complementary strands, is the key to its unique capacities. DNA directs the creation of new cells and, since it is present in the eggs and sperm, the reproductive cells, it also transmits hereditary information from one generation to the next.

Each DNA strand is like one side of a ladder, each rung of which consists of a pair of chemical units known as bases. There are four kinds of these bases, or half-rungs: adenine (A), thymine (T), guanine (G), and cytosine (C). A joins up only with T, and G only with C; thus, the structure of one side of the ladder specifies the structure of the other side. DNA replicates itself by "unzipping" the long strings of joined bases, and by assembling new complementary strands under the supervision of a suite of chemicals known as enzymes. Errors in this assembly process are very rare—about one in a billion—and sometimes have most undesirable results; yet they also provide a source of variation between parents and their offspring that is vitally important to the evolutionary process.

Working in conjunction with its close relative ribonucleic acid (RNA), DNA also directs the construction of proteins, the basic "building blocks" of living organisms. It is from proteins that are built all the various tissues of our bodies, ranging from bone to brain. The units of which the proteins themselves are made are known as amino acids, and twenty different kinds of them find their way into our cells as the result of the breakdown of food. These raw materials are assembled into proteins according to instructions carried in the sequence of bases along the DNA strand. The bases work exactly as the letters of the alphabet do: three bases in line act like a three-letter word which specifies a particular amino acid.

DNA at work. This diagram shows how the chainlike structure of the complex DNA molecule allows it to replicate itself. The two halves of the spiraling "ladder" which makes up the DNA molecule have "rungs" made up of one of four bases: adenine (A), guanine (G), cytosine (C), and thymine (T). To form the complete ladder, A joins only with T, and G with C. Thus, when the ladder is "unzipped" down the middle, each side of the ladder exactly specifies the structure of the other half, which is reassembled from new molecules available within the cell. DNA also directs the building of proteins, the molecules which perform most of the cell's chemical activities. Its intermediary in this process is its cousin RNA. RNA is similar to DNA, but is single-stranded (a "half-ladder"), and it substitutes the base uracil (U) for T. RNA is built up using the sequence of DNA bases as a template, and it transfers this information by acting itself as a template for the production of proteins. Proteins are long chains of amino acids, molecules available in the cell as a result of the breakdown of food. Each sequence of three RNA bases acts like a "word" specifying a particular amino acid, and thus a length of DNA, equivalent to a sentence, contains all of the instructions needed for assembling a complete protein.

1.

2.

3.

4.

5.

When the structure of DNA was first unraveled in the 1950s, it had been known for many years that hereditary information is carried in discrete units, known as genes, which are found within the reproductive cells of the parents. As early as 1866 the Moravian Abbott Gregor Mendel (1822–84), experimenting with flowering pea plants in his monastery garden, had been able to show that offspring do not simply represent a "blend" of their parents' characteristics, but that genes remain intact, albeit reshuffled, from one generation to the next. DNA itself was identified as that hereditary material by the bacteriologist O.T. Avery in the 1940s. But the discovery of DNA's structure by Watson and Crick was the essential prelude to breaking the "genetic code" and to all the later advances that have brought us today's sophisticated view of what genes are and how they function.

It's been discovered, for instance, that not all segments of DNA are responsible for specifying particular protein molecules. Many do this, of course, and these are known as structural genes. But other bits of DNA appear to be involved in the regulation of biological processes via the turning on and off of other genes, and are thus known as regulatory genes. Yet other sequences of bases serve to punctuate the genetic language. What's most remarkable, however, is that vast parts of the "genome" (all of your DNA) seem to be "silent," serving no apparent function at all. What we seem to have here is the residue of millions of generations of evolution. Apparently the evolutionary process works as much by making parts of the genome redundant as by modifying existing DNA or adding to it.

Figuring out how DNA works has ultimately given birth to the exciting new science of biotechnology, which holds enormous promise for the future. Biotechnologists use a variety of ingenious tricks to make genes and to introduce them into cells. Once inside, the gene tells the cell to make proteins which were not produced before, and this offers enormous potential benefits to medicine, in addition to such enticing prospects as better-tasting tomatoes. Up to now, these benefits have been more or less

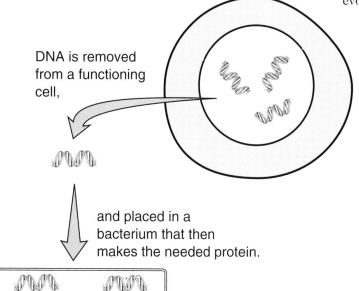

DNA is removed from a functioning cell,

and placed in a bacterium that then makes the needed protein.

Next, the protein is purified,

and finally injected into the patient.

The major steps of basic genetic engineering, in which genetically altered bacteria are used as "factories" to produce substances needed by patients who lack the genes to make them. Advanced techniques of "gene therapy" offer the prospect of introducing genes directly into patients' cells to restore their function.

Illustration by Diana Salles

indirect, as when biotechnologists have altered very simple one-celled organisms to produce vast quantities of medicines needed by human patients. These include insulin, used in the treatment of diabetics. But in the future the methods of genetic engineering may be applied more directly to humans suffering from genetic disorders.

Medical geneticists have described several thousands of inherited diseases which are caused by improperly functioning genes in the cells of the body. Cystic fibrosis, for example, is a relatively common disease in which excessive secretions within the lung cause progressive respiratory failure. This happens when a particular gene specifies a faulty form of a protein called CFTR; this problem protein disrupts the flow of electrically charged molecules in and out of the cells that line the lung. If the right form of the gene could be introduced into the cells of patients with cystic fibrosis, the condition potentially could be alleviated. Something similar has, in fact, already happened in the case of a much rarer condition called Severe Combined Immunodeficiency Disease (SCID), which leaves patients defenseless against all sorts of infections. Sufferers from SCID lack the gene that in the rest of us tells our cells to make an immune-system protein called ADA. Recently, researchers at the National Institutes of Health in Bethesda, Maryland put copies of the ADA gene into the white blood cells of a young SCID patient and saw an improvement in her condition.

One of the big problems in biotechnology is, of course, how to get improved genes into human cells. Obvious candidates for this job are the viruses, which consist of threads of DNA or RNA in a simple protein coat. Viruses cannot replicate outside of living cells, and thus cannot strictly be considered as living themselves. Nonetheless, they are phenomenally successful as parasites on living beings, as epidemics from the common cold to AIDS dramatically attest. Viruses are expert at penetrating cells and changing the way they function to their own benefit; if they could be "infected" with particular genes, they might in turn be able to infect the body cells whose functioning doctors wish to improve. Many researchers are concentrating their attention on this possibility.

Primitive life forms

If we cannot consider the extremely simple viruses to be alive in the strictest sense, what are the most primitive life forms on Earth? Here we have to turn to the bacteria. Bacteria are organisms whose DNA is not walled off from the rest of the cell in a nucleus surrounded by a membrane; and although bacteria are typically single-celled, they are amazingly diverse in habitat and function. This is hardly surprising when one considers that bacteria have been around longer than any other life form: fossil bacteria have been found that are 3.5 billion years old.

Indeed, it is to the action of early bacteria that we owe the conditions that allowed the emergence of more complex life forms, for the Earth's early atmosphere was very different from the oxygen-containing mixture of gases that we depend on today. The atmosphere coalesced originally from gases that separated out from the Earth's solid components after the globe began to form about 4.5 billion years ago. Initially it probably consisted mostly of nitrogen, carbon dioxide, and water vapor, plus a variety of other gases such as hydrogen, methane, and ammonia. The earliest bacteria probably built themselves from substances that arose in this atmosphere, deriving their energy through breaking these compounds down by a fermentation-like process.

It was the next innovation of the bacteria, however, that provided the real breakthrough. This was photosynthesis, a process well-known today in plants, which manufacture sugar, starch, and other carbohydrates from water and carbon dioxide with the aid of the energy in sunlight. We owe most of the subsequent increase in diversity and complexity of living forms to those photosynthesizing bacteria of 3 billion years ago. Here's why.

Part of a stromatolite, a layered structure built by billions of microscopic blue-green bacteria living together in a shallow sea and trapping sediment among them. The earliest stromatolites are about 3 billion years old, and examples are still forming today in Shark Bay, western Australia. This particular example comes from rocks in Death Valley about 1.5 billion years old.

Photo by
Willard Whitson.

The primitive photosynthesis carried out by the early bacteria most likely differed from that of plants because bacteria used hydrogen, carbon dioxide, and carbon-based organic molecules; but quite quickly the primitive bacteria were joined by "blue-green" bacteria that used water as a base material, and which produced large quantities of oxygen as a byproduct.

Billions of these tiny organisms living together typically produced large layered structures known as stromatolites (literally, "coverlet stones," on account of this layering like bedclothes). The earliest of these appear in the fossil record at about 3 billion years ago. As photosynthesizing organisms of this kind multiplied, increasing levels of oxygen appeared in the atmosphere and this eventually made respiration possible. A form of energy production based on the breakdown of oxygen, respiration permitted an eighteen-fold increase in energy yield, and an oxygen-rich atmosphere opened up a whole new set of possibilities for life forms on Earth.

Some bacteria of the primitive kind remain with us today. And since the longer a group of organisms evolves, the more variation is likely to occur, this is one reason for the claim that, at least in regard to metabolism (the chemical processes that take place within the cell), the differences found among the tiny bacteria are larger than any among the other living forms of the world, even those as different as humans and trees.

In the modern world, bacteria are amazingly successful and are still the dominant life forms, at least in terms of sheer numbers. In a single lump of soil the size of a golf ball there may be 10 billion of them, and they are found everywhere, particularly in and on animals of all kinds. In our daily lives we most commonly hear about bacteria in the context of human diseases (or cures—many antibiotics are produced by bacteria). But much more importantly, we depend on bacteria for many of our basic life processes. For besides producing much of the oxygen we breathe and otherwise influencing the concentrations of gases in the atmosphere, bacteria promote many important biological reactions, perhaps most notably those involved in the digestion of the food we eat.

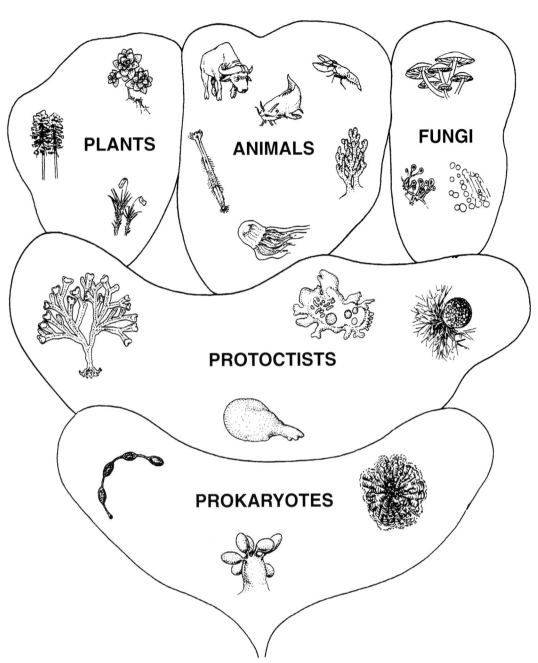

The "five kingdoms" of life (discussed on the next page). Because of the extraordinary diversity of living things, and of the enormous amount of change that has accumulated in the various lineages over evolutionary time, it has proven very difficult to arrive at a satisfactory classification of the major groups of living organisms. The diagram above, based broadly on the "five kingdom" arrangement first proposed by R.H. Whittaker in 1959 and more recently modified by Lynn Margulis and colleagues, attempts to show how these five major groups fit together in terms of relationships and function. The vertical axis is not strictly one of time, but rather one of increasing complexity—which does, however, indirectly reflect the passage of time. As noted in the main text, the protoctists are a "wastebasket" group, whose interrelationships are unresolved; however, the kingdom Protoctista does fill a useful function in bridging the gap between the bacteria and the three other eukaryote kingdoms (plants, animals and fungi), among which relationships are also currently unclear. *Illustration by Diana Salles.*

The many kingdoms of life

Living organisms come in a vast array of shapes and sizes, from bacteria and mushrooms to the sequoia tree and the blue whale. It's thus hardly surprising that one of the toughest questions in the entire study of biology concerns how to classify this broad array of life forms into major groups. For as scientists have gradually learned more about the extraordinary variety of living things, they have realized that the simple traditional division of life forms into animals and plants is inadequate.

From the mid-nineteenth century on, for example, scientists have become increasingly aware that bacteria and fungi are at least as different from each other as are plants and animals. More major groups clearly needed to be recognized, but which, and how

The Eukaryotic Cell. The cells of eukaryotes are incredibly tiny, mostly under one ten-thousandth of an inch in diameter. Yet each one is a complete factory, containing many different "machines" within its enclosing membrane. This membrane selectively allows passage of materials in and out of the cell, which is filled with a gelatinous *cytoplasm* in which the other structures float. The cytoplasm is divided up by the *endoplasmic reticulum*, a network of membranes on which *ribosomes* are found. These are sites where amino acids, formed from the breakdown of food, are assembled into proteins, the "building blocks" of life. The cytoplasm also contains such structures as the *Golgi bodies*, which modify and repackage proteins, and the *mitochondria*, which provide the cell's energy by breaking down organic molecules that enter it. The mitochondria have their own DNA and divide independently of the rest of the cell, suggesting that these were once independent bacteria-like entities that were "captured" by early eukaryotic cells. The nucleus is walled off from the rest of the cell by a double membrane, and within it is found the rest of the cell's DNA as well as the *nucleolus*, where the ribosomes are made. Various other structures also occur within cells, depending on their function. In the human body alone there are more than 200 different kinds of cells, each allowing the tissues they comprise to perform different tasks. This illustration shows a generalized eukaryotic cell, with some of its contained organelles.

Illustration by Diana Salles.

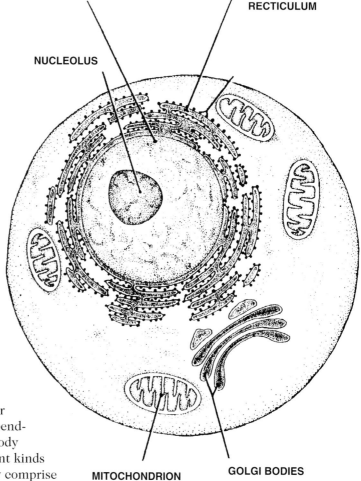

NUCLEUS

ENDOPLASMIC RECTICULUM

NUCLEOLUS

MITOCHONDRION

GOLGI BODIES

many? This continues to be a tricky issue, for the very variety of the world's living things tends to obscure the similarities that might be used to recognize how they are interrelated. But although there is no unanimity on the matter, many scientists have come to recognize five major groups—or "kingdoms"—of living things, while recognizing that at least one of these groups may not be "natural" in the sense that all of its members are descended from the same exclusive common ancestor.

The first of these five kingdoms is **Prokaryotae**. This contains the bacteria, which we have already seen are organisms in which the DNA is not separated from the other components of the cell in a well-defined central "nucleus" bounded by a membrane. Indeed, the name Prokaryotae simply means "pre-nucleus." All of the other living forms, ourselves included, are known in contrast as **eukaryotes** ("true nucleus"); and, whether they are unicellular or multicelled as we are, they all have more complex cells of the kind described in the neighboring box.

The first eukaryotes to appear on Earth were single-celled, and their fossils have turned up in rocks that are as much as 1.3 billion years old. It is widely believed that such early eukaryotes originated when various different prokaryote ancestors came together to live within the same cell membrane, each benefiting from the different abilities of the others. Combinations of this kind may in fact have come about more than once, giving rise to a number of distinct lineages of eukaryotes. We've already looked briefly at the prokaryotic bacteria; now let's look at the four kingdoms of eukaryotes.

Seaweeds and slime molds: "wastebasket" organisms

The kingdom **Protoctista** is a convenient "wastebasket" for a vast variety of organisms—most, but not all, single-celled—that are otherwise hard to classify. Its members are neither bacteria, nor fungi, nor animals, nor plants. One of the key features that distinguishes between the major groups of multicellular organisms is the nature of the clusters of cells from which individuals develop. Such primordial clusters vary greatly: among fungi they are known as spores, among plants as embryos, and among animals as blastulas. These different cluster types are highly distinctive; and protoctists are alike in developing from none of them. All protoctists are, of course, eukaryotes, with characteristically eukaryotic cell structure and organization. But apart from this they have little in common, except that all live in watery media: the seas, fresh waters, or the fluids contained within other organisms.

Without being clearly related to the plants, fungi or animals, various protoctists have some of the characteristics of one or more of these groups. Predictably, then, their lifeways are enormously varied. Some protoctists are single-celled, others multicelled; some photosynthesize, others feed (some do both); some are parasitic, others are free-living; some actively move, others are passive.

The simplest protoctists are the caryoblasteas, giant cells which have a membrane-bounded nucleus but lack most of the other features typical of eukaryotes. It's because of this that they are often considered to be the most primitive of eukaryotes. More typical amoebas and a host of other single-celled creatures are also considered protoctists, as are the slime molds in which amoeba-like cells aggregate to form a single mass. And while most protoctists are microorganisms, the largest of them include the kelps and seaweeds, huge algae that can grow to hundreds of feet in length. Because the group is so varied, knowing that an organism is a protoctist doesn't tell us much at all about its relationships or adaptations.

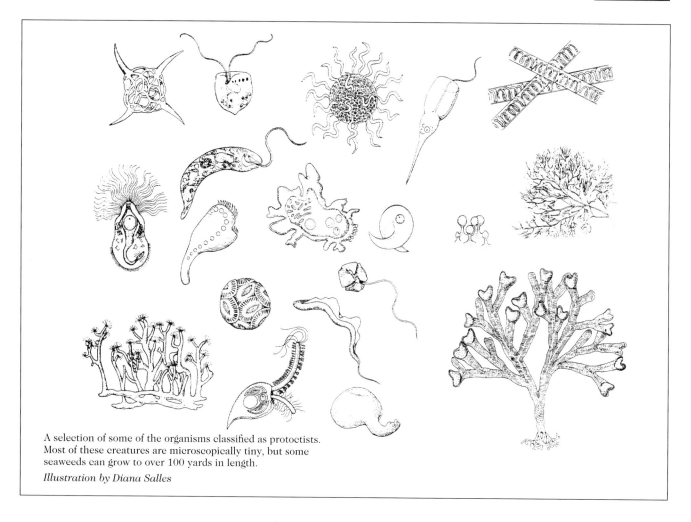

A selection of some of the organisms classified as protoctists. Most of these creatures are microscopically tiny, but some seaweeds can grow to over 100 yards in length.

Illustration by Diana Salles

Fungi: giants of the forest

Some fungi are among the very largest and longest-lived organisms known. In 1992 a group of mycologists showed that enormous masses of filaments embedded in the floor of a Michigan forest all belong to the same individual of the species *Armillaria bulbosa*. This single individual fungus occupies an area of over forty acres of forest floor, and probably weighs in total over one hundred tons. This is comparable to the weight of a large blue whale; and although this figure is dwarfed by the one thousand tons of the largest sequoia trees, the latter owe most of their weight to nonliving wood, while the fungus is composed of active tissues. This remarkable organism is believed to have been growing for at least fifteen hundred years, and its edges are still expanding outward at a rate of over six inches per year.

Members of the kingdom **Fungi** have been called the "garbage disposal units" of the forest. This is because these filamentous organisms feed by injecting enzymes into the organic materials on which they grow, breaking them down into nutrients. This action places them among nature's principal agents of decomposition, the process by which organic materials are recycled.

The major "garbage disposers" are the saprophytes. These dispose of hard, woody substances such as fallen trees and prevent the buildup of dead materials that would otherwise eventually suffocate the growing plants. Saprophytes are responsible for most of the 85 billion tons of carbon dioxide released into the atmosphere each year. One of the most surprising recent discoveries about saprophytes is that some of them actively "hunt" the microorganisms that are found abundantly in their habitats, luring and trapping their tiny live prey in a variety of ingenious ways. Among their main victims are the minute nematode worms, found in huge numbers wherever the soil is covered with vegetation, as well as tiny water-loving organisms called rotifers and a variety of amoebas and bacteria. Once the victim has been detected and trapped by the fungus, it is often paralyzed by toxins and its body is invaded by filaments and consumed from within.

The other major group of fungi, the mycorrhiza, actively promote the growth of living plants, with which they have an intimate ecological relationship. Their filaments form sheaths around tree roots, and make water and minerals available to them with great efficiency. From the trees they obtain sugars produced in the leaves. Pines and other coniferous trees are totally dependent on the fungi for their nutrients, and birches and some other broadleaved species will also die without the active help of mycorrhizal fungi. This symbiosis is fundamental not only to individual tree species, but to the overall ecology of the forest.

From the human point of view, fungi are often inconveniently responsible for disease and decay. But at the same time they do offer some life-enhancing attributes. Mushrooms and truffles, for instance, are the fruit bodies of a variety of fungal species, and their main job is to carry the spores that give rise to new generations of fungi. But they are also prized for their well-known role on the table as delicacies, and some of them have been revered in certain cultures for their provision of psychedelic experiences. Furthermore, it is the activities of the unicellular fungi known as yeasts that make possible the production of cheese and wine and which, among other things, make bread rise.

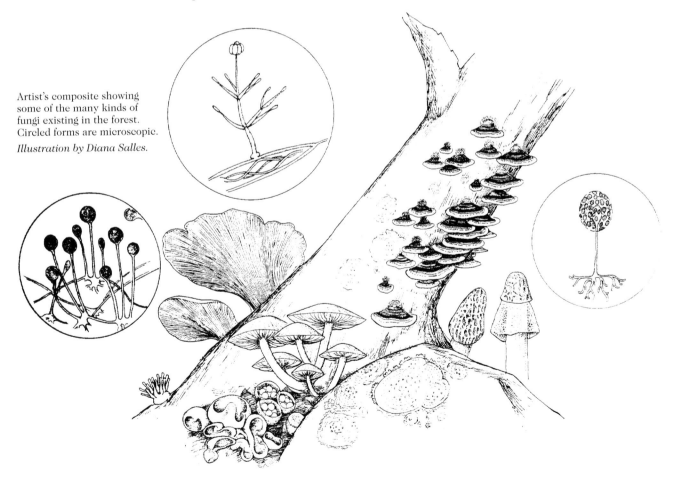

Artist's composite showing some of the many kinds of fungi existing in the forest. Circled forms are microscopic.
Illustration by Diana Salles.

Plants: energy from sunlight

Probably the simplest way of describing members of the kingdom **Plantae** is as eukaryotes that photosynthesize. This means that they use the energy in sunlight to convert carbon dioxide and water to carbohydrates and oxygen. A tiny minority of plants has, however, lost the ability to photosynthesize, so to find the characteristic that unites all plants today we have to look at the unique and specialized mode of early development that all share.

Plants have been the predominant feature of the terrestrial landscape throughout the 430 million years of their fossil record. Further, they exist in an astonishing variety of forms and play numerous roles in the working of the Earth's ecosystem. The most important such role is the fixation of carbon, the conversion of atmospheric carbon dioxide into the carbon-based molecules of which living organisms are composed. In the process, they incidentally give off the oxygen on which respiration depends. Besides plants, only algae and photosynthetic bacteria perform these services.

The result of this is that, at least on land, plants are the "primary producers" of the ecosystem, providing the carbon compounds which other organisms need to live. This places plants at the base of the food chains on which complex ecologies depend (although prokaryotes are also active in this role on land, and are overwhelmingly the major players in the seas). Plants are fed on directly by primary consumers such as herbivorous mammals. And the herbivores in turn are eaten by secondary consumers, carnivores such as ourselves. At the same time, the decomposers break down the detritus of plants, both kinds of consumers, and other decomposers. Each kind of actor has a role in the complex ecological play; but on land it is plants that furnish the basic element around which the script revolves.

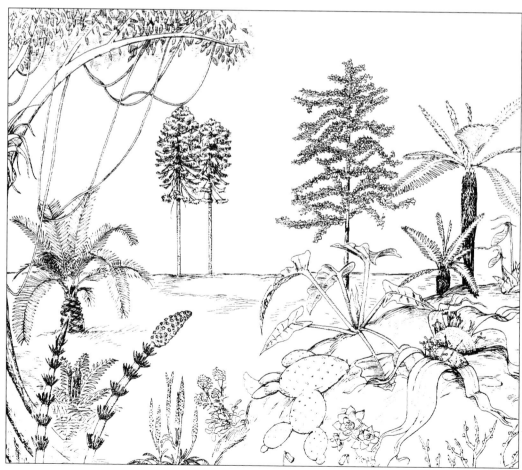

Composite scene showing some of the many thousands of plant species alive today.

Illustration by Diana Salles.

Artist's view showing some of the vast diversity of animal species alive today.
Illustration by Diana Salles.

Animals: enormous variety

The kingdom **Animalia** includes organisms as dissimilar as sponges and whales. Animals are all multicellular (what we used to call "unicellular animals" are now classified as protoctists), developing from blastulas, hollow balls of cells derived by sexual reproduction. All animals are also heterotrophs ("eating others"), which is to say that none of them is a primary producer. There are more than thirty major animal groups, or phyla, recognized in the animal kingdom, in addition to about sixty phyla of other living organisms. And humans form the merest minute fraction of a subphylum of only one of them, the subphylum Vertebrata. Viewed in this light, we occupy only the tiniest twig on the great branching tree of animal life.

Animals exhibit an astonishing diversity, from aquatic worms to birds. Yet of the thirty-plus animal phyla, only two have invaded the land with any conspicuous success. One of these is Arthropoda, the group that contains the millipedes, insects, scorpions, etc. The other is Chordata, our own group. Most chordates are backboned animals, otherwise known as vertebrates, but the chordate net is actually cast wider than this.

For to be a chordate you have to have three defining characteristics, at least at some stage of your development as an individual. These characteristics have been inherited by living chordates from the extremely remote common chordate ancestor, and perhaps the most important of them is the presence of a nerve cord along the upper portion of the back. It is this structure that in vertebrates has become the brain and spinal cord, and it is stiffened by the second feature, known as the notochord. This is a flexible rod that lies between the nerve cord and the gut, at least early in development. It was around the notochord that the backbone developed in the earliest vertebrates. The third characteristic is the presence, again at least during early development, of gill slits in the throat. Gill slits may have been first acquired as feeding organs, but gills themselves are best-known as the structures that allow fish to extract oxygen from water; and it's a tip-off to our marine ancestry that we humans have them at all, even though they are lost very early in the prenatal life of each individual.

These three features are the principal reason why we and the other vertebrates belong to the same phylum as the small marine organisms known as the tunicates and lancelets. Adult tunicates spend their lives anchored to a rock, but their larvae are free-swimming with notochords etc. For their part, the lancelets are small, primitive, vaguely torpedo-shaped aquatic creatures that apart from their chordate credentials lack much to get excited about.

The backboned animals

With the vertebrates, however, we begin to find the principal elements of our familiar body plan, notably a backbone containing a spinal cord that extends behind a skull containing the brain. But it's only in one subgroup of the vertebrates that we find a structure with which we can really identify: one with four limbs and an elaborate air-breathing apparatus. This is Tetrapoda ("four feet"), the group of vertebrates whose ancestors left the oceans and invaded the land about 350 million years ago, 150 million years after the first vertebrates appeared in the seas. Their descendants comprise the amphibians (frogs and salamanders), reptiles (from turtles to dinosaurs), and mammals (shrews, elephants, whales, birds, etc.).

The amphibians, however, are still tied to life in the water to the extent that they have to lay their eggs and fertilize them there, and they spend their initial period of development in that medium. It was the ancestor of the reptiles, birds and mammals that took the next step in the process of emancipation from water. This innovation was the development of the so-called amniote egg, enclosed by an amnion, a membrane that contains water as well as the nutrients needed by the developing embryo. Having their own water supply, fertilized amniote eggs can survive on land, breaking the cycle of dependence on a watery environment in which to live. In the next chapter we will look more closely at the vertebrates as exemplified by one of their amniote species: *Homo sapiens*.

Chapter Two
Humans Are Vertebrates

Vertebrates are animals with a backbone. They include three classes of fish and four classes of mainly land-living tetrapods: the amphibians, reptiles, birds, and mammals. Although humans share the same essential structure and body systems as fish, the necessities of living in the water are very different from those of living on land. Only the other tetrapods have body structures that closely resemble our own. Of course, there are many differences among tetrapods in the details of these body structures and systems, but by and large they are reasonably similar.

All tetrapods—and indeed all vertebrates—have an internal skeleton and the muscular system needed to move it. They have sensory organs and a nervous system to gather, process, and respond to information from the environment, and to coordinate the workings of the body. They have respiratory and circulatory systems to extract oxygen from the environment and to circulate it to the tissues along with nutrients extracted by a digestive system. They also have an endocrine system that regulates growth and the functioning of the organs of the body. All of these systems (except for the respiratory system of fishes, which extracts oxygen not from the air but from water), are sufficiently similar among the vertebrates to be understood by reference to one particular vertebrate, *Homo sapiens*. In the rest of this chapter we will look at the structure and functioning of this particular "higher" vertebrate.

The skeleton and muscles

Hard though it is, the bone that composes the human skeleton is a living tissue that constantly responds to stress and performs many functions beyond simply providing the basic framework of the body. For example, it also serves as a reservoir of minerals, and new red blood cells are produced in its marrow.

Most human bones are preformed in the flexible substance known as cartilage. This is gradually transformed into bone in each one of us as we develop from embryo to adult. As we mature, new bone is laid down in the growth centers of the bones. Some bones are composed of more than one growth center; these are separated by cartilaginous areas that grow together to produce a single bone as development is completed. The internal structure of bones is formed of a strong type of protein fiber called collagen. On this flexible organic base are deposited calcium, apatite, and other minerals which give the structure rigidity. The mineral content of bone varies according to the amount of stress placed upon the skeleton. Mineral deposition decreases where stress is low, as among weightless astronauts and the bedridden, while it rises in highly active people such as athletes.

As the central supporting structure of the body and its organs, the skeleton is constructed broadly according to principles familiar to engineers. Each bone is shaped to withstand the various stresses with which it must cope. For example, since the thigh bone must sustain forces that tend both to bend it and to compress it, its shaft is formed as a hollow tube, which provides the optimum ratio of strength to weight. But because our basic structure is inherited from a remote ancestor with a very different lifestyle, our design is not perfect; an engineer who could start with a blank piece of paper could probably do better.

For example, we are descended from precursors who bore their weight on four limbs. In such quadrupeds, the backbone is flexed rather like an archer's bow, and it absorbs stresses by bending and straightening as a unit. The weight of the body is supported at four points, and the internal organs are suspended below the spine much like the roadway of a suspension bridge. To achieve upright posture, we have thrown our vertebral column backwards and upwards, placing the entire weight of the trunk above the pelvis. To permit this our spine has acquired a double curve, a dangerous thing for a structure that is not only in compression but which is made up of many bones separated by cartilaginous pads (discs). Many of our back problems, slipped and ruptured discs in particular, may be traced to our having had to adapt a quadrupedal body to an upright lifestyle.

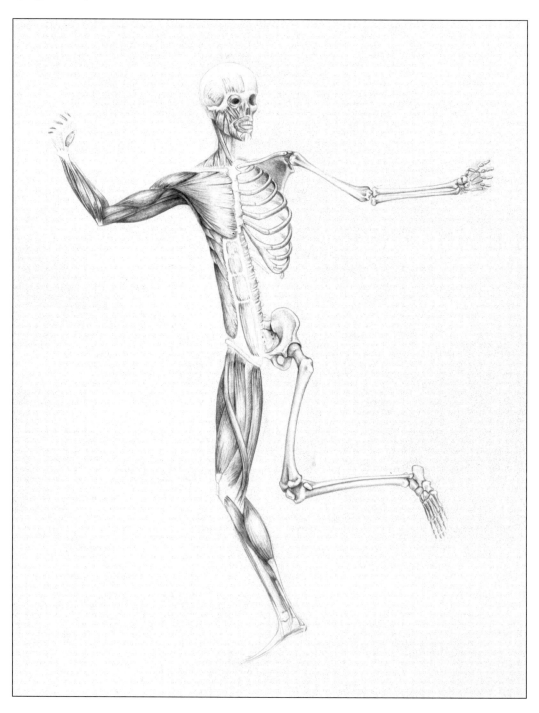

The framework of the human body is our skeleton, a structure of rigid bony elements that meet at movable joints. Movement of the joints is produced by the contraction of muscles across them.

Illustration by Christine Rossi.

Bones come together at unions known as joints. Some joints move freely, others very little. Some joints, such as the knee joint, are like hinges, permitting free movement in only one direction. Ball-and-socket joints like the hip allow movement in all three planes; but since every joint represents a compromise between stability and mobility, ball-and-socket joints are less stable than hinge joints. In other kinds of joints, surfaces slide across each other.

Moving joint surfaces are lined with cartilage, which varies in type according to the kind of load the joint has to bear. Like all mobile joints, mechanical and biological alike, those of the skeleton require lubrication. This is achieved by enclosing the joint in a sac that contains a specialized lubricant known as synovial fluid. The combination of fluid and cartilage results in minimal friction between the opposing surfaces. The integrity of the joints themselves is assured by ligaments, tough fibrous bands that surround the joint capsules and bind the opposing bones together.

Between the joints the bones of the skeleton are rigid, which means that they can act like a system of levers. These levers are moved by muscles that stretch across the joints from one bone to the next, and that contract to achieve bone movement. Muscles may attach directly into the bone, or their ends may taper into tough tendons which can extend for some distance before inserting into the bone. The muscles that move the bones of the skeleton are of a special type. Known as skeletal muscles, they consist of bundles of long fibers. On command from the nerves that control them, these fiber bundles contract (shorten) and relax (lengthen). Some fibers contract fast while others do so more slowly, but more steadily; muscles vary in the proportion of each kind of fiber they contain, depending on the function they serve. Most skeletal muscles or muscle groups act in pairs, one member of the pair contracting to flex the joint and relaxing while it extends under the contraction of its antagonist. When they contract together they act to stabilize the joint in a given position.

Digestion

Digestion is the process whereby food is reduced to molecules that can be sent to the various tissues of the body and used as a source of energy. Among vertebrates, this process takes place in a long tube (about twenty-five feet in humans) that runs from the mouth to the anus. Our digestive tract is thus technically outside the body proper, which can be considered as the walls of the tube; nutrients enter the body cavity by absorption through the gut.

The initial phase of digestion takes place in the mouth by chewing. The breakdown of starches is begun by an enzyme in the saliva, of which we produce copious amounts, and at the same time our molar teeth grind the food mechanically into smaller particles. After being swallowed, the food passes down the throat to the esophagus; the muscular outer lining of this tube then contracts in waves to propel the food to the stomach. At the base of the esophagus is a muscular ring known as the cardiac sphincter, which contracts to prevent irritating stomach acids from flowing back toward the mouth. If the sphincter does not close down properly, or if it opens inappropriately for some reason, we experience the distressing reflux of stomach acids that we call heartburn.

The stomach itself is an elastic pouch that lies under the diaphragm. Its inner walls are lined with cells that produce hydrochloric acid and a precursor that turns into the digestive enzyme pepsin when in contact with the acid. The "smooth" muscle of the stomach walls contracts to aid further mechanical breakdown of the food, while the acid and digestive enzymes act to reduce it chemically.

The human digestive system consists in essence of a single long tube within which nutrients are absorbed from the food, and from which waste products are eliminated. Its main components are the esophagus, stomach, and small and large intestines.

Illustration by Christine Rossi.

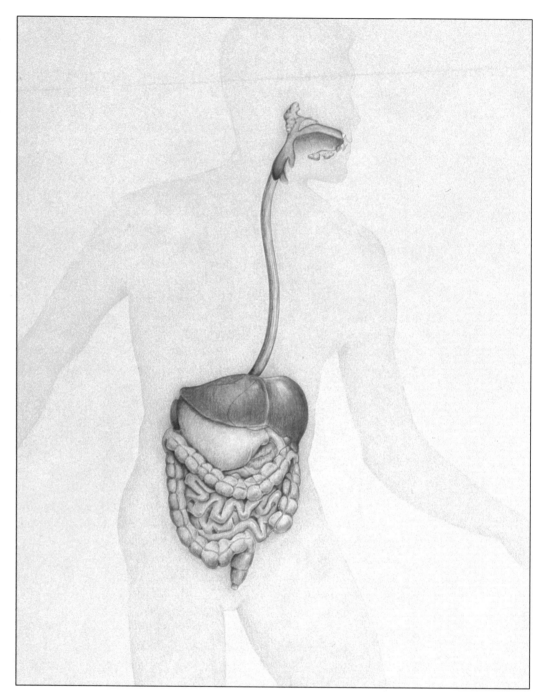

Beyond the stomach lies the small intestine. Defying its name, this tube is over 20 feet long in the average human and is divided into three parts: the duodenum, jejunum, and ileum. The inner walls of the small intestine are complexly folded to increase its surface area, and from these walls project millions of tiny structures called villi. These villi separate nutritive particles of food, such as proteins and sugars, from substances such as cellulose that the body cannot break down. The villi have an ample supply of blood and very thin walls, so nutrients, minerals, water, and other needed items can easily pass through them and into the bloodstream. The blood then carries them to the liver and on to the various tissues of the body.

The small particle size required for passage through the villi is achieved when a variety of substances attack the partly digested food as it enters the small intestine. Such substances include bile (from the liver) that breaks down fats and further digestive enzymes from the pancreas. This organ also contributes antacids to neutralize the stomach acid still present. The end result of this chemical attack is to break down the nutritious components of the food into simpler compounds that the body can use. Proteins are reduced to amino acids, carbohydrates to glucose, and so forth.

From the small intestine, what's left of the food passes into the large intestine, so named because it is much wider, though much shorter, than the small intestine. The large intestine is also divided into three sections: the caecum, colon, and rectum. Here water is separated from the waste materials and reabsorbed through the intestinal walls, and resident bacteria synthesize various needed substances. Most of the approximately twenty hours that the average plateful of food passes in the human digestive tract is spent in the large intestine, before the waste products are finally expelled through the anus.

"Chemical messengers"

The endocrine system regulates most of our bodily functions, including our growth, sexual development, reproductive function, fluid balance, metabolism, and responses to stress. It consists of a network of glands which are linked by hormone "messengers." Hormones are specialized chemical substances which the endocrine glands release into the bloodstream. They are then transported to the "target" tissues on which they act.

The endocrine system is an extremely complex one in which perfect balance must be maintained, for deficiencies or excesses in hormone production can cause serious disorders. For example, insulin, secreted by the pancreas, regulates the body's metabolic processes and the uptake of nutrients, glucose in particular. If the pancreas produces too little insulin, diabetes results, while too much insulin produces low blood sugar. Similarly, deficiencies in thyroid hormone production result in lethargy, while an overactive thyroid produces a rapid heartbeat and an increased metabolic rate that gives rise to weight loss and muscle weakness, as well as to general nervousness.

Much of the endocrine system is controlled by the pituitary gland, which is situated under the base of the brain. It secretes a large number of different hormones, most of which stimulate either growth or the functioning of other glands. Since its products influence so many other glands, the pituitary is often called the "master gland" of the body. However, the pituitary is itself closely regulated by the hypothalamus, the part of the brain under which it lies. It is thus the hypothalamus that is the true master gland, and besides controlling the pituitary, it releases several hormones directly into the bloodstream.

The endocrine system is, however, very far from being a one-way affair, with the brain and endocrine glands simply sending chemical commands to the target tissues via the bloodstream; in fact, it depends totally on an intricate "feedback" relationship between the endocrine organs and the tissues that respond to their signals. For example, it is the level of glucose in the blood that tells the pancreas when to secrete insulin, and it is sex hormones from the ovary that signal the pituitary gland and the brain when an egg is ready to be released during a woman's menstrual cycle. But simple levels of hormones and other chemicals in the blood are not the whole story. Many hormones are secreted not continuously but in pulses whose pattern conveys messages to the target tissues. Such pulsatile hormone secretion is extremely important in the female menstrual cycle and in the process of sexual maturation.

Scientists have been able to isolate and purify many of the hormones of the endocrine system, so that people with abnormalities of their endocrine glands can usually

The endocrine system consists of a series of glands that release hormone "chemical messengers" into the bloodstream under the control of a part of the brain known as the hypothalamus. The hypothalamus mainly acts via the pituitary gland, which lies below it. Various other glands are seen in this illustration, including the adrenals on the kidneys.

Illustration by Christine Rossi.

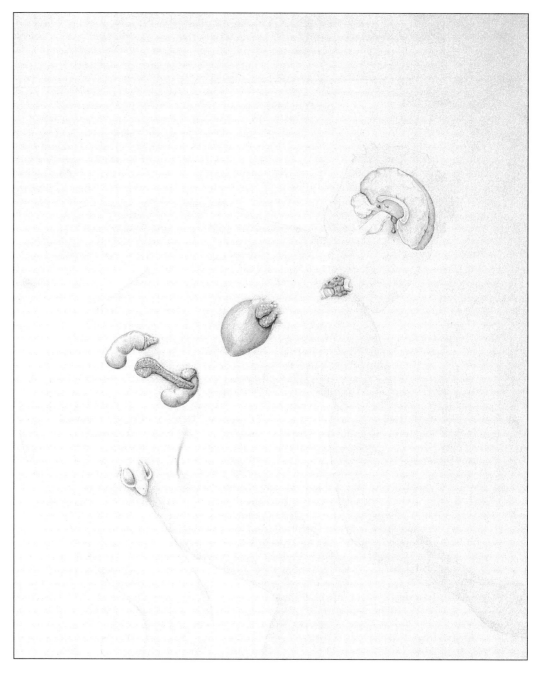

be successfully treated by a physician who fully understands the complex workings of the system. The classic example of such treatment is, of course, the administration of insulin to diabetics, which was made possible by the isolation of this hormone back in the 1920s. Since then, research has greatly expanded the possibilities. For example, gonadotropin-releasing hormone can be given to help restore fertility in women who do not produce this hormone naturally. Many human hormones have already been identified, and more are being identified all the time. Indeed, such structures as the intestines also are turning out to be endocrine organs, producing a variety of hormones that are involved in the digestion of food. Intriguingly, many of these "gut" hormones are found in the brain, where their function is still not understood.

Breathing

The body needs oxygen to convert food to energy. The job of the respiratory system is to extract oxygen from the air and to remove carbon dioxide, one of the waste products that forms in the body during the process of energy conversion.

The central element in the respiratory system is the lungs. These are large, spongy air sacs into which air is drawn through the nose and mouth via the windpipe, or trachea. This flexible tube divides below the larynx (voicebox) into two smaller tubes known as bronchi, one going to each lung.

Of course, food and air are taken in through the mouth and throat. In most mammals, though, including newborn humans, the air and food passages are quite separate. The larynx is high in the throat, where it is able to lock into the space at the back of the nasal cavity (the nasopharynx). In this position, the larynx provides a direct passageway for air between the nose and the lungs, so food can pass to either side of the interlocked larynx and nasopharynx, and on to the stomach, without interrupting breathing. But in adult humans the larynx is much lower in the throat. The food and air passages cross above the larynx, creating a situation where simultaneous breathing and swallowing are impossible. This arrangement also makes it possible for a lump of food to lodge in the entrance to the larynx, blocking the airway.

The advantage of this potentially disastrous arrangement is that it enormously expands the volume of the pharynx that lies above the voicebox. This is crucial in producing articulate speech, because production of necessary sounds depends on our unique capacity to modify the resonance of air in this large chamber. The respiratory system in humans thus contributes powerfully to our ability to speak, one of mankind's most singular attributes.

The lungs divide into separate units called bronchopulmonary segments. Each of these is supplied with air by smaller air tubes (bronchioles) that stem from the bronchi and ultimately terminate in clusters of tiny air sacs, or alveoli. Each of the millions of alveoli is enclosed in a network of tiny blood vessels so small that only one red blood cell at a time can pass through. Blood returning from the tissues is laden with carbon dioxide and depleted of oxygen; on its way through the alveoli, its red cells pick up new oxygen molecules which diffuse through the alveolar walls. At the same time, carbon dioxide diffuses out of the plasma, the clear liquid in which the red cells travel, and into the air spaces of the lungs. The newly oxygenated blood then travels to the heart, whence it is pumped around the body, and the carbon dioxide is breathed out when the next breath of air is expelled from the lungs.

Air is breathed into the lungs when the diaphragm muscle below them contracts and forces the bottom of the thoracic cavity down. Other muscles move the rib cage up, and atmospheric pressure forces air into the lungs to fill the enlarged cavity. When the diaphragm relaxes, the lung cavity returns to its normal size and squeezes air out of the lungs. The quantity of air moved in and out of the lungs in this way each day amounts to some three thousand gallons. Pure air contains about 20 percent oxygen. About a fifth of the oxygen in the air is normally absorbed during breathing, while air breathed out usually contains about four times as much carbon dioxide as the air breathed in. Although we can consciously control our breathing rate if we wish to, the activity of breathing is essentially under unconscious control by the brain. Breathing in is automatically stopped when the lungs are fully expanded, while the breathing rate is raised when oxygen consumption is increased, as reflected by rising carbon dioxide levels in the blood.

The lungs and the diaphragm below them form the heart of the body's respiratory system. In the lungs waste gases are removed from the blood and replaced by the oxygen needed by the tissues. In humans the airways above the lungs provide the sounds associated with speech.

Illustration by Christine Rossi.

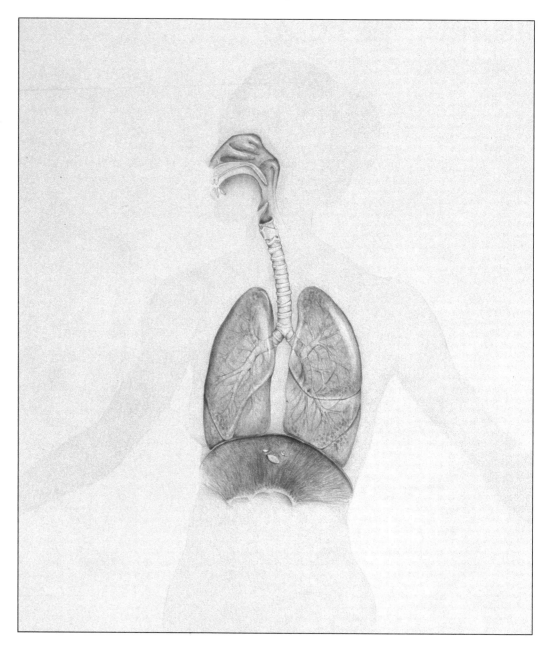

Circulation

The bloodstream is the body's transportation system. To assure proper body functioning, the tissues require oxygen and nutrients and must dispose of waste products; hormones must travel between glands and organs; antibodies must get to where they are needed to fight infection and disease; and internal temperature must be maintained. The circulatory system is essential to all of these functions.

The center of the circulatory system is the heart, which powers the circulation of blood throughout the body. It is a large, muscular organ which rhythmically contracts to pump oxygenated blood away from itself through the tough, muscular-walled arteries. The residual pressure assures the return to the heart of the depleted blood through a system of veins. These blood vessels are less thick-walled than the arteries, but contain valves to prevent the blood backing up. When a person is resting, the heart recirculates all of the body's blood in about one minute; during energetic activity, it greatly speeds up circulation. The heart pumps constantly. It's been estimated that each day the average individual's heart pumps about two thousand gallons of blood through some six thousand miles of vessels of all sizes. This is a lot of recirculation for the body's five quarts of blood.

In mammals, the heart, which is constructed of a very specialized type of muscle with its own inbuilt rhythm of contraction, is divided into four spaces called chambers. The thinner-walled chambers above, called atria, receive incoming blood; the thicker-walled ventricles below pump blood out. Depleted blood arrives in the right atrium and continues to the right ventricle, where contraction forces it out along the pulmonary arteries to the lungs. From the lungs, oxygenated blood returns to the left atrium via the pulmonary veins and is passed along to the left ventricle, whence it is pumped back to

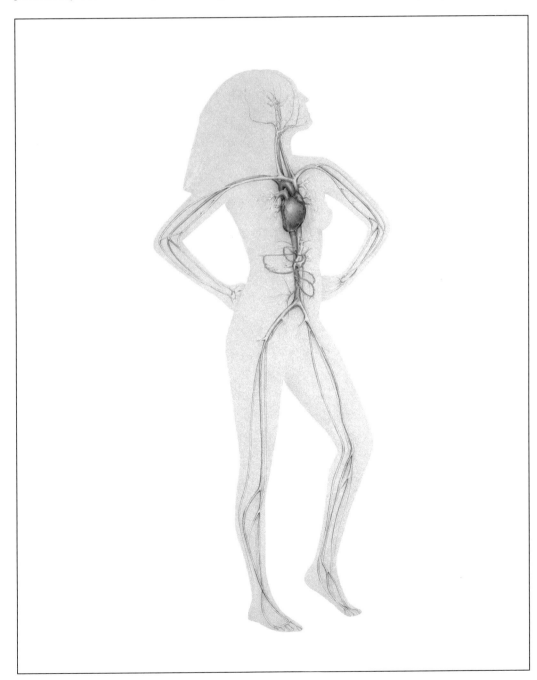

Circulation of the blood throughout the body is powered by the heart muscle. A system of arteries carries the blood away from the heart and to the tissues; blood returns to the heart via a system of veins. The blood passes from the right hand side to the left-hand side of the heart via the lungs, where it picks up oxygen and waste gases are removed.

Illustration by Christine Rossi.

the body tissues through the arterial system. A system of valves prevents backflow while the heart muscle is relaxed.

As the main arteries course further away from the heart, they divide into smaller and smaller vessels until they terminate in the exceedingly tiny tubes known as capillaries. Through the thin capillary walls, oxygen and nutrients diffuse from the blood to the tisses, and waste products travel the other way. The capillaries eventually join up with the system of veins, which coalesce to form larger vessels the closer they get to the heart. When the depleted blood reaches the heart, it is dispatched to the lungs, and the cycle starts over.

Since the heart contracts and relaxes rhythmically, the pressure of the blood in the blood vessels varies. The two-figure measurement reported by physicians consists of a

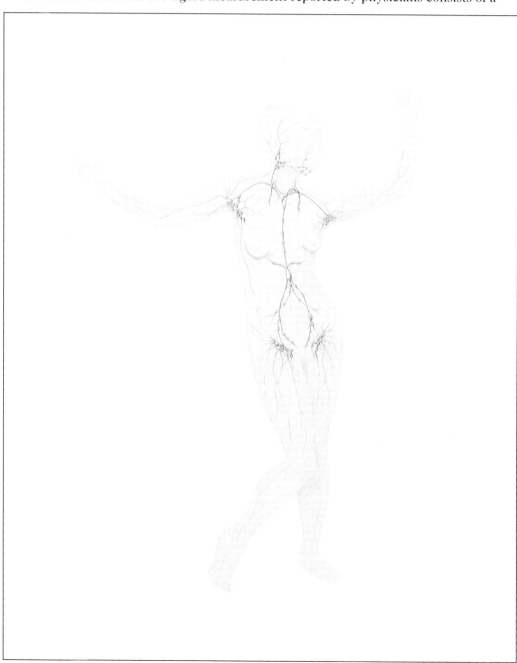

The lymph system is a secondary circulatory network that collects unwanted particles and fluids from the interstices of the body. Lymphatic vessels travel from the tissues to points known as "nodes," and on to the venous system of the bloodstream.

Illustration by Christine Rossi.

higher number known as the "systolic" pressure, as the ventricles contract to send a new rush of blood out through the arteries, and a lower figure known as the "diastolic" pressure, measured in the momentary space between contractions. Blood pressure can be raised by various undesirable conditions that cause either an increase in blood volume or excessive resistance to the movement of blood through the vessels. Arteries can also become clogged, starving body organs of needed fresh blood. When this happens in one of the "coronary" arteries that supply the heart itself with blood, the result is a heart attack. When it occurs in one of the arteries supplying the brain, the consequence is a stroke. These are both serious events because both the heart muscle and brain tissue are extremely sensitive to oxygen deprivation.

The blood itself is a remarkable fluid. It consists principally of red cells, white cells and platelets, all suspended in a clear fluid called plasma. The red cells, as we've seen, transport oxygen to the tissues. The many varieties of white cells fight infection, while the platelets perform an invaluable role in blood clotting and the repair of damaged blood vessels. The plasma is the vehicle in which these different kinds of cells travel, but it also has its own role in transporting hormones around the body, and in conveying nutrients to the tissues and waste products away from them.

Another important function of blood is transport of the heat generated by chemical processes in the cells of the body. In mammals it is critically important in maintaining a constant body temperature. In hot conditions the blood carries metabolic heat from the depths of the body to its surface, where it can be lost to the outside; in cold temperatures, on the other hand, the capillaries near the body's surface constrict, minimizing such heat loss. Some mammals even have specialized networks of blood vessels that transfer heat from outgoing to incoming blood when conditions are cold.

One final aspect of circulation in the body is the lymphatic system. This is in a sense a secondary circulation system, whose functions include collecting unwanted materials from the tissues of the body and removing excess fluids from intercellular spaces. It also serves as an important defense system against infection. Lymphatic vessels travel from the tissues, through filtering points known as nodes, and collect into larger ducts that eventually drain into the system of veins. The waste products are then disposed of along with those scavenged by the blood. The lymph system's major role in fighting foreign bodies gives it a particular clinical significance. The swelling of lymph nodes is often one of the clearest signs of infection.

The body's cleaning mechanism

The urinary system is the major cleansing system of the body. As the tissues convert food to energy, waste products are formed that must be disposed of. The lungs dispose of carbon dioxide; the liver detoxifies the blood; and the urinary system deals with most other waste products. In addition, though, the urinary system also controls fluid balance within the body. This is an important function, since the average adult human contains well over ten gallons of water within and between the cells of his or her body. This quantity must be regulated as fluid is lost on the one hand by sweating, respiration, and excretion, and gained on the other through drinking and eating.

The chief components of the urinary system are the kidneys, which are the major filtering organs, and the bladder, or collecting sac. Tubes connect the kidneys to the bladder, and the bladder to the outside of the body. Every day, some five hundred gallons of blood flow through the twin kidneys, situated on either side of the spine at the base of the rib cage. Blood enters each kidney through a major artery that, once inside, divides into numerous tiny capillaries. Minute bundles of these vessels, known as glomeruli, are connected to tiny tubes, or tubules, the whole forming a structure called a nephron.

The major component of the urinary system is the kidneys, which filter unwanted products from the blood. Fluid containing these waste particles flows to the bladder, a holding sac, from which it is eliminated directly to the exterior.

Illustration by Christine Rossi.

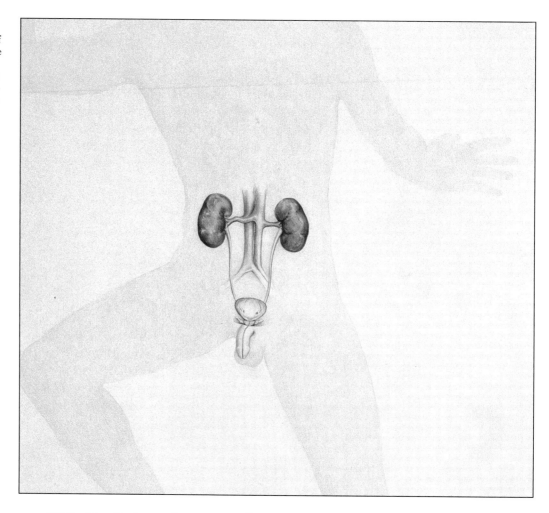

While blood laden with impurities flows through the nephron, large particles such as blood cells stay within the glomerulus and continue in the bloodstream. Smaller particles such as salts and some fluids, however, pass through its walls and into the tubules. From the tubules, water and some useful products are selectively absorbed back into the surrounding capillaries, while fluids containing unwanted substances collect into progressively larger ducts which ultimately leave the kidneys and connect with the bladder.

The selective reabsorption that goes on in the tubules is a very delicate process, governed by several hormones in a complex feedback monitoring system, and is critical to maintaining the correct composition of body fluids. It is also a very conservative system. On average, only about 1 percent of the large quantity of fluid recycled by the kidneys each day finds its way to the bladder to be excreted as urine. The amount of fluid excreted is, of course, closely controlled by the kidneys according to external conditions and the body's internal needs. On hot days, for instance, we cool ourselves by copious sweating, and thus need to excrete as little fluid as possible; on colder days, we need to lose more water by direct excretion.

The urinary system is completed by the bladder, a sac with strong, muscular walls that lies in the hollow of the bony pelvis. It is the holding point for the urine that drains from the kidneys through twin tubes called ureters. A single tube called the urethra connects the bladder to the outside, and through it urine is evacuated from the body. In male humans, but not in females, the urethra also forms part of the genital tract.

Tiny stones composed of calcium compounds or of uric acid can form at any point from the kidneys downstream, distending the various tubes and causing considerable pain. As the principal disposer of nongaseous metabolic wastes, the urinary system can be helpful, however, in diagnosing and monitoring metabolic diseases. This is done by analyzing the composition of the urine it produces. In normal individuals, for example, blood glucose is almost entirely reabsorbed by the kidneys. But in diabetics blood glucose levels are too high, and the kidneys simply cannot absorb it all. The excess appears in the urine, where its presence indicates that sugar metabolism is amiss. The famous thirst that accompanies diabetes is caused by the need for extra fluid in which to excrete the glucose.

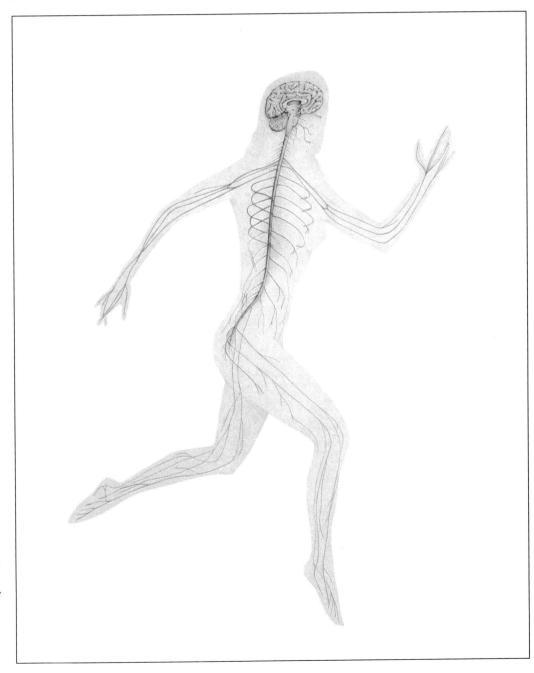

The brain is the body's control system, and it communicates with the rest of the body by a system of nerves. Some of these nerves, such as those to the eyes, emerge directly from the brain; others travel via the spinal cord, branching off to distant parts of the body.

Illustration by Christine Rossi.

The nervous system

The nervous system is the master control and coordination network of the body. It consists of the brain and spinal cord, plus the nerves that feed information to and from these central structures. The brain not only governs our thoughts and feelings, it also controls most of the body's activities, including both voluntary ones like walking and involuntary ones such as breathing. From both the brain and the spinal cord, motor nerves carry messages to structures such as the sense organs, the muscles, and the skin, while "sensory" nerves feed information back from them to the brain.

These twin types of nerve form the somatic nervous system, which coordinates our responses to information that comes in from the environment. A second, autonomic, nervous system communicates between the brain and the internal organs of the body that operate outside our conscious control, regulating such functions as breathing and digestion.

The autonomic nervous system has two subsystems, the sympathetic and parasympathetic. These operate in opposition to each other, the sympathetic subsystem stimulating the organs and the parasympathetic slowing them down. Regulation of autonomic function thus consists of finely balancing the effects of the two subsystems. In some cases, responses to stimuli are so automatic that the spinal cord responds without consulting the brain. If you pick up a hot plate the message only has to get as far as the spinal cord before you reflexively drop the offending item as fast as possible.

All nervous-system functions originate in communication between individual nerve cells, or neurons. The sites of such communication are the synapses, where a "chemical messenger" emitted by one cell stimulates activity in another. Within the neuron, such impulses are transmitted as electrical signals. These travel inward along short projections called dendrites, which may communicate with several other cells, and outward again along one long projection known as an axon to the next neuron. In this way nerves, which consist of bundles of neurons whose axons may in certain cases be a yard long, can almost instantaneously transmit signals originating in the brain to distant organs, or sensations received by the body back to the brain.

The brain itself is an extremely complex organ, with a layered structure consisting of different components that have been successively acquired or elaborated in our lineage since the time of our remotest vertebrate ancestors. The most primitive part of the brain is the brain stem, which is really the enlarged top end of the spinal cord. Different parts of the brain stem control such basic functions as normal breathing and heartbeat, and emit and receive somatic nerve impulses that communicate with the muscles and skin. One center of the brain stem also controls our sleep-wake cycle, and stimulates higher centers of the brain to activity.

Behind the brain stem lies the cerebellum, which is mostly concerned with bodily coordination. It is of critical importance, for instance, in receiving input from the various muscles and muscle groups and in synchronizing their activities to produce smooth movement, as in walking or running. It also receives direct input from the organs of balance in the middle ear, synthesizes it with other information received, and thus assures the body's equilibrium.

By far the largest part of the human brain is the region called the cerebrum, which overlies the brain stem and cerebellum. It is divided into left and right halves, or hemispheres, which communicate via a bridge called the corpus callosum. Broadly, each hemisphere of the brain controls the opposite side of the body. This explains why a stroke (interruption of blood flow) on the right side of the brain can cause paralysis of the left side of the body. Extensive wrinkling of the outside of the human brain reflects the great expansion of the cerebral cortex, the "grey matter" which overlies the "white matter" of the underlying cerebrum. The cerebral cortex controls our voluntary body movement and much else besides. Particularly distinctive in humans is the expansion of the "association areas" of the cortex. These permit close communication among the various specialized areas of the cortex (which communicate separately, for instance, with the various organs of sense) and are ultimately responsible for our distinctive intelligence.

Beneath the cortex lies a group of cerebral structures loosely called the limbic system. These control emotions such as fear and anger, and basic body drives such as sex and thirst. Adjacent to the limbic structures we also find the thalamus, which transmits sensory information to the cortex, and the hypothalamus, the part of the brain that controls the endocrine system.

As the discussion above implies, we know quite a lot about the structure of the brain and even about the functions of many of its constituent parts. But as yet it is not at all clear how its various parts work together to produce that peculiar quality we know as consciousness.

Our senses

Five senses supply the brain with information about the environment: sight, smell, taste, hearing, and touch. A sixth sense, balance, tells the brain about the body's orientation in space.

Vision

Although many organisms get along perfectly well with limited vision or even no vision at all, visual information is extremely important to humans. By one estimate, 80 percent of the information processed by our brains enters through the eyes. We see because the sun emits energy rays that are reflected off objects in the environment. The eyes convert some of these rays—the visible spectrum—into nervous impulses that travel to the visual center of the brain for interpretation. Thus the brain, rather than the eyes themselves, actually "sees" the world around us.

Each eye is a roughly spherical structure which is protected at the front by a transparent layer called the cornea. Behind the cornea is the iris, the pigmented circle that makes your eyes blue or green or brown and which opens and closes to control the amount of light that enters the eye. Light passing through the pupil, the hole in the center of the iris, is focused by the lens on the retina, which occupies the hemispherical surface at the back of the eyeball. The lens is flexible, and muscles pull on it to focus the light properly according to the distance of the object being viewed. The space between the lens and the retina is filled by a clear gelatinous substance that helps to maintain the eyeball's spherical shape.

The retina is densely packed with two kinds of receptors, rods and cones, that are sensitive to light. Rods respond acutely to light intensity but provide only monochrome vision. Cones are color-sensitive, one type to each primary color, but they require higher light levels to function properly. Rods vastly outnumber cones even in the human retina (some species have no cones at all), but cones monopolize the center of our retina where incoming light is concentrated. This gives us exquisite color vision during the day but places us at a disadvantage in twilight and darker situations, relative to mammals that retain all-rod retinas.

A layer of nerve cells at the back of the retina converts the arriving light stimuli into nerve impulses, which it passes on to the optic nerve. This in turn sends them along, via a way station, to the primary visual cortex at the back of the brain. At a structure called the optic chiasma, the optic nerves from the two eyes meet. Some of the fibers from the left eye continue from there to the right side of the brain, and vice versa, so that impulses from the visual fields of both eyes end up in the same half of the brain. This forms the basis for depth vision, which is achieved by comparing the images received from the left and right eyes.

Humans are among the most highly vision-dependent of all vertebrates, and we are thus unlike our early ancestors who probably relied mostly on smell and lacked color vision. Indeed, the early evolutionary story of the group to which we belong was very largely the history of the triumph of vision over smell.

Smell and Taste

Although our sense of smell has now become overshadowed by other senses, notably vision, we are more influenced by it than we think. Both anatomically and chemically, taste is closely related to smell. Many of the subtleties that we perceive as taste are actually detected by smell, and both senses depend on chemoreception, or the recognition of certain types of molecules.

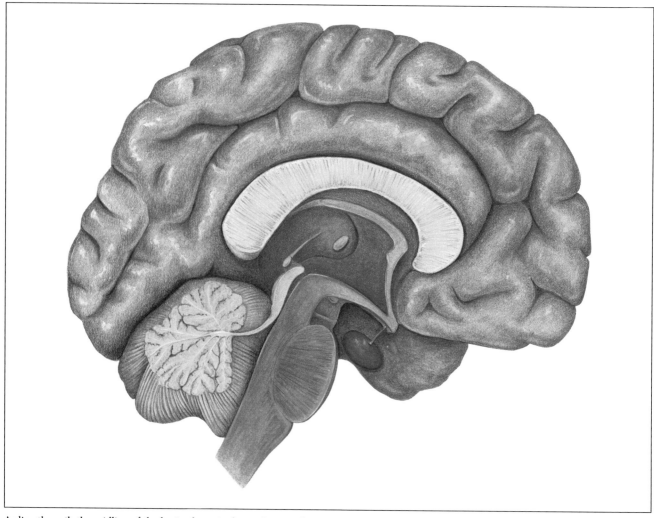

A slice through the midline of the brain shows its layered internal structure. At the bottom is the "brain stem," which continues as the spinal cord and which controls many "involuntary" actions such as normal breathing. Behind it lies the coordination center known as the cerebellum. The upper portion of the brain is called the cerebrum. In humans the cerebrum is extensively wrinkled to enlarge its surface area, and it is in this structure that "higher" functions such as memory and cognition take place.

Illustration by Christine Rossi.

The taste receptors on the tongue and parts of the palate and throat detect only four basic sensations: sour, sweet, salt, and bitter. They also have very local distributions. The sweet receptors are the furthest forward on the tongue, followed by the salt receptors, which are flanked by those for sour and followed by the ones for bitter. This is one reason why wine tasters are careful to swirl the liquid around the insides of their mouths before swallowing it, so that all of the taste receptors will be stimulated. The other reason is that this action volatilizes chemicals in the wine so that they enter the air in the oral cavity and can be carried up to the smell chemoreceptors at the top of the nasal cavity. These receptors form a small patch that communicates directly with the olfactory part of the brain, beneath the frontal lobes. This patch is covered in mucus, in which the molecules that convey smell dissolve before they are identified. Both taste and smell are obviously of great importance in analyzing foods that are eaten, but in many mammals, smell also plays an important role in communication, such as individuals leaving behind them scent "marks" that can be detected by others who come along later.

Hearing

Sounds are vibrations transmitted through the air to the ear and sent along to the brain to be deciphered. The mammal ear itself consists of three parts: external, middle, and inner. The external ear gathers sound vibrations from the air and directs them down the tube that connects the eardrum to the outside. The eardrum itself is a membrane that is stretched across the inner end of this tube, and which vibrates with the air in it. On the inside of the eardrum, a set of three tiny bones amplifies these vibrations and transmits them across the middle ear space to the "oval window" at the entrance to the inner ear. A part of the innermost bone fits snugly in this opening, and its vibrations cause pressure changes in the fluid inside the snail-shaped cochlea of the inner ear. These pressure

This cross-section through a human eye shows its major components as well as the long optic nerve that communicates with the brain. Incoming light rays pass through the pupil, which expands and contracts according to their intensity, and are collected by the hemispherical retina at the back of the eye after being focused through the lens. _Illustration by Christine Rossi._

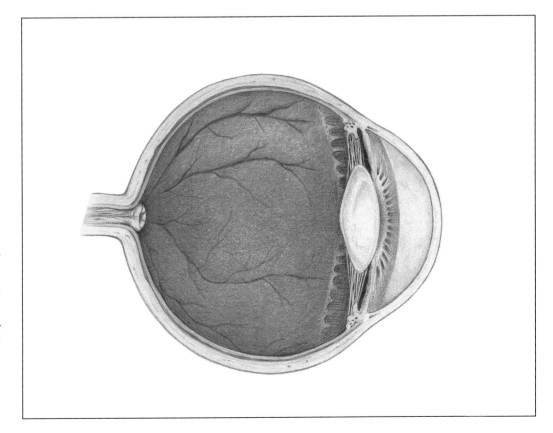

changes are converted by ultrasensitive "hair cells" into nervous impulses which travel to the brain where they are interpreted as sounds.

The middle ear is an enclosed space, and the air inside it would interfere with hearing if its pressure became too low or high. To equalize the pressure in the middle ear with the air outside, the pharyngotympanic (Eustachian) tube runs between the middle ear and the back of the nasal cavity. This is an essential function, of course, but it's also a mixed blessing because it provides a way for infection to spread extremely rapidly from the nose and throat to the ear region. Discomfort because of Eustachian tube problems is particularly common in young children, in whom the tube is straighter than in adults.

The range of sounds that humans are able to detect is in fact relatively limited and declines with age. Many mammals can detect much higher frequency sounds than we do (remember the "silent" dog whistle?), and it is possible that our ears have become specialized for efficient reception of the range of sounds produced in normal speech.

Balance

The inner ear also contains three semicircular canals, each of which is oriented in a different dimension. Each fluid-filled canal is lined with sensitive hair cells. These detect movements in the fluid and convey this information to nerve fibers, which ultimately pass it on to the cerebellum. The three-way orientation of the semicircular canals ensures that movements in all directions are detected. The canals constitute our principal organ of balance, although in most of us vision also plays a role in maintaining our body equilibrium. Indeed, motion sickness results at least in part from the brain's reception of conflicting signals from the ears and eyes.

Touch

Our skin, the largest organ of the body, serves not only to protect our internal structures but to receive sensations of contact from the outside world that are transmitted to the brain through a rich supply of nerve endings. These touch receptors, which are all terminal branches of various somatic sensory nerves, exist in dazzling variety—as you might expect, given the extraordinary array of sensations that impinge on us. Not all touch receptors are in the skin itself; the pains in arthritic joints, for example, are picked up deep below the skin by nerve endings of this same general class.

Touch is possibly more fundamental to us both as social and biological beings than any other sense. Humans, like all social primates, depend on touch more than we generally realize to develop and maintain social bonds, from infancy onward; at the other end of the spectrum, pain is a warning that allows us to avoid dangerous situations or exertions before we are injured.

Chapter Three
Humans Are Mammals

One of the first things any aspiring evolutionary biologist is taught is not to look at the spectrum of living vertebrates as a *scala naturae*, a progression from the "lowest" to the "highest," with *Homo sapiens* at the pinnacle. Each species, we're sternly told, is the product of its own unique evolutionary history, and ours and every other species is like a twig at the extremity of a branching bush rather than like a rung of a vertical ladder. But it's only with a major effort that any of us manages to dismiss the notion of the *scala naturae* entirely, for it fits so neatly with much of the order that we perceive in nature. This is because certain groups of organisms have changed less over time than others have. For example, not all fish are equally related to us, but all conserve the body proportions of the (aquatic) ancestral vertebrate more closely than we do. And modern crocodiles, while having a long pedigree, have changed impressively less in the time since their cousins the dinosaurs than our own lineage has. We thus tend to think of them as "less evolved."

Because such differences in the rate of change have had highly visible results, the *scala naturae* idea has had no difficulty in surviving, and it particularly continues to affect some of the terminology we use. For example, among the vertebrates the two "warm-blooded" classes, Aves (the birds) and Mammalia (the mammals), are traditionally reckoned to have achieved "higher vertebrate" status, particularly in having acquired sophisticated physiological mechanisms that permit maintaining body temperature independent of environmental conditions. Being thus endothermic allows these animals nocturnal as well as daytime activity and makes it possible for them to live in a much wider range of environments.

Endothermy does all this, however, at the cost of amazingly high energy consumption: a reptile can get by on less than one-tenth of the food required by a mammal of the same size. For reptiles, this is extremely handy in some hot but impoverished environments such as deserts, where behavioral adaptations can compensate for limitations of physiology; but the behavioral flexibility offered by endothermy is generally reckoned to give the mammals an edge under most circumstances.

Beyond the fact that they share endothermy, however, birds and mammals don't have many specializations in common except perhaps that parents of both often invest a lot of energy in raising their offspring, and that both have circulatory systems in which four-chambered hearts provide separate circulation for oxygenated and depleted blood. These similarities, however, were probably independently acquired: birds—accurately described as "feathered dinosaurs"—have a line of descent from reptiles that is quite independent of the one that gave rise to mammals, and differ from them in many fundamental ways.

Mammals are distinctive above all in that mothers nourish their offspring with milk produced in their mammary glands, and in having a skin that is covered with hair, at least at some stage of the life cycle. Like feathers, hair is derived from reptilian scales, and is probably a specialized structure related to endothermy. From the rather prosaic initial function of preventing loss of body heat, hair has acquired all kinds of social and other significances in the lives of mammals. Although some mammals have to one extent or another lost body hair over the long history of this remarkable vertebrate class (the earliest mammal fossils known are well over 200 million years old), this doesn't belie their underlying unity. Nonetheless, living mammals are an exceptionally diverse group, and despite the fact that all share milk production, it is their reproductive systems that provide the main evidence for dividing them into three major subgroups. Let's briefly look at these three basic kinds of mammals: the monotremes, marsupials, and placentals.

Egg-laying mammals

The egg-laying mammals are called monotremes. Among all the great array of living mammals, only the Australasian spiny anteaters and platypuses survive to represent this group. Both kinds of monotreme have quite specialized ways of life; but in laying leathery-shelled eggs similar to those of reptiles, they may recall the reproductive pattern of the very first mammals.

Young monotremes hatch from the egg while they are still relatively undeveloped. Among the spiny anteaters, the egg hatches within a pouch that forms temporarily on the mother's belly; in the platypus, incubation of the egg(s) and early development of the offspring takes place within a burrow to which the mother remains largely confined. In both cases, the young receive nourishment from milk patches on the mother's skin. These patches are formed of sweat glands that have been modified to produce milk. In contrast to other mammals, the milk ducts don't come together in localized nipples but exude milk evenly through the specialized skin of the patch. The monotremes have a mammal-like internal temperature-control system, though body temperature does vary more than in other mammals, and is also a bit lower.

In addition to egg-laying, the monotremes share various other attributes with reptiles including a number of features of the skeleton and the sharing by the reproductive, digestive and urinary tracts of a single external orifice, the cloaca. However monotremes do not simply represent a stage of development that is somehow intermediate between reptiles and other mammals; the monotremes are in fact highly specialized in ways of their own. For example, both spiny anteater and platypus adults lack teeth, a basic feature of most mammals. Whereas all of the teeth of reptiles are simple, pointed, and replaced continuously as they wear, mammals have four distinct types of teeth which are replaced only once (as when, in our case, the "milk teeth" are pushed out by the adult teeth). In the front of a mammal's mouth are the incisor teeth, used to cut and crop food; behind these are the pointed canines, followed by

The echidna or spiny anteater, *Tachyglossus*, is one of the two surviving genera of monotremes, the egg-laying mammals.

Illustration by Diana Salles.

the premolars and the molars, which reduce the ingested food. The upper and lower teeth fit precisely against each other, and there is usually a sideways component in jaw movement that permits a certain grinding action. Among the astonishingly diverse array of living mammals, all four kinds of teeth vary greatly in shape, size, and number, and they tell us a great deal about the kinds of foods a particular species eats. In spiny anteaters, however, teeth have been dispensed with entirely: a long and roughly tubular jaw system occupied by a long, sticky tongue serves to take in ants, worms, and so forth. Teeth appear in young platypuses, but they are rapidly replaced by the famous leathery "duck bill" that facilitates their aquatic feeding. It's difficult to see exactly how—given their unique skull structure— the monotremes are related to the other mammals.

The pouched mammals

The "pouched" mammals are called marsupials. These animals are most familiar to North Americans in the form of the opossum, but they exist in the greatest variety in Australia and New Guinea, where the kangaroos are probably their best-known representatives. Marsupials give birth to live young, but at a very early stage of development—sometimes only a little over a week after fertilization. At birth, the tiny infant marsupial struggles to a protective "marsupial pouch" formed on the mother's belly. Inside the pouch it finds a nipple, on which it suckles until it is ready to explore the outside world. But even when it has ventured forth from the pouch, a young marsupial will dart back inside whenever alarmed.

There are a number of unique specializations associated with the pouch-rearing of young. For example, if the pouch is occupied, the development of a fertilized egg can be suspended until pouch space is available for the new offspring. On the other hand, marsupials are primitive in certain respects: like monotremes they have a single external orifice for reproduction and the elimination of solid and liquid wastes.

In Australia the marsupials diversified enormously. It's been generally thought that this was because they didn't have to compete with placental mammals until humans arived on the continent, but recent discoveries of apparent early placental fossils may throw some doubt on this. However this may be, what is truly amazing is how closely the diversification of different kinds of Australasian marsupials paralleled the diversification of the placentals elsewhere in the world. Among the marsupials there are not only forms that mimic all the major activities of

Kangaroos are among the best-known of the marsupial mammals, among which the young are born at a very early stage and complete most of their development in the mother's "pouch." Inside the pouch they are nourished by milk from the mother's mammary glands.

Illustration by Diana Salles.

placentals—gliding, leaping, hunting, burrowing, swimming, and so forth—but there are Australian marsupials which look remarkably similar to their placental counterparts in other regions. This observation is probably the most dramatic evidence anywhere that adaptation to specific habitats really does occur in nature.

This is not to say that functional differences do not exist. Marsupials, for instance, lack a corpus callosum, the brain structure that in placentals allows the two cerebral hemispheres to talk to each other. But marsupials are similar to placentals in all the ways in which monotremes are in being quite efficient thermoregulators, in having four different kinds of teeth, and in other characteristics as well as in giving live birth. Yet, in general, marsupials haven't competed very successfully with placentals, either in the New World (as shown by the fossil record) or in Australia, where many a variety of marsupial has become extinct following colonization by humans and their placental fellow travelers. One exception to prove the rule is, of course, the opossum, whose successful invasion of North America is continuing.

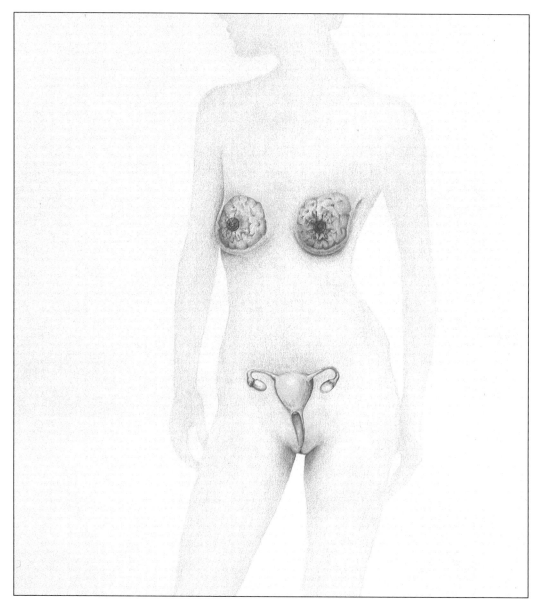

The human female reproductive system. Eggs are produced in the ovaries, travel along the uterine tubes, and, if fertilized, implant into the wall of the uterus, or womb. Within the uterus the embryo is nourished by the placenta, which allows nutrients to pass from, and waste products to enter, the mother's bloodstream. After birth, the infant is nourished by milk from the mother's mammary glands.

Illustration by Christine Rossi.

The placental mammals

The placentals are overwhelmingly the most successful group of mammals, having diversified in the 65 million years since the extinction of the dinosaurs to produce forms as different as bats, whales, shrews, elephants, rats, mongooses, and humans. Placental mammals are found in every habitable environment on Earth, from the frigid deserts of the polar regions to the Amazonian rain forest to every ocean. They eat grass, plankton, meat, fruit, insects, flowers. They run, swim, fly, burrow, leap, scurry. They live alone, in families, in vast herds. They depend on sight, or on sound, or on smell. They have flippers, hooves, claws, wings, or grasping hands. They have large brains, or small ones. They are active by day, by night, by twilight or around the clock.

Among all placental mammals, the young are born at a relatively advanced stage of development. This is permitted by the placenta, a structure that begins to form in the wall of the mother's womb when implantation of a fertilized egg occurs. The developing fetus is connected to the placenta by the umbilical cord, through which the fetal blood flows. The placenta is also in intimate contact with the mother's blood circulation. The fetal and maternal blood supplies don't mix, but oxygen and nutrients diffuse from the mother to the fetus through the membranes of the placenta, while waste products from the fetus go the other way. This exchange provides a highly efficient method for meeting the increasing needs of the fetus over a long developmental period.

The process that leads to the implantation in the uterus of the fertilized egg begins when the ovaries of the female release one or more ripened eggs (ova). In humans, only one egg is normally completely ripened and released per cycle. Each ovary is adjacent to the end of a uterine ("Fallopian") tube, which collects the egg and transports it toward the uterus. In the tube it becomes available for fertilization by sperm produced in the testes of the male and introduced into the female's vagina, to which the uterus is connected. Only one sperm can fertilize the egg, but many are needed to dissolve its protective coat. Fertilization occurs when the DNA of the egg (a complete cell) unites with that of the sperm (which is little more than DNA in a protein coat). Following successful fertilization, the egg, or zygote, travels down into the uterus and implants itself into the prepared uterine wall. The highly complex process of development then follows.

Gestation lengths vary greatly among placental mammals, from about two to three weeks in mice and shrews, to twenty-two months in the African elephant. Clearly there's a size effect here, but longer gestation times also are typical of precocial mammals, whose offspring are few and relatively well developed at birth and tend to receive a lot of parental attention. Shorter gestation times, in contrast, are usual among altricial mammals, which have higher reproductive turnovers and less well-developed offspring, and invest less parental effort in each one. Following birth, placental infants are nourished with mother's milk. And, as among marsupials, the milk ducts are grouped at highly localized nipples. The nipples may range in number from one to several pairs, altricial species tending to have more.

Humans are primates

Our own tiny corner of the enormous variety of the placental mammals is occupied by the order Primates. This group dates from the dawn of the Age of Mammals, well over 60 million years ago, and is plainly precocial, setting the stage for brain size increase. It also almost certainly had its origin in a forest setting. The name "Primates," which simply means "the first," was coined by the eighteenth-century Swedish savant Carolus Linnaeus (1707-1778), the founder of the modern system by which we classify living things. In Linnaeus's day, of course, virtually nobody in the scientific establishment doubted the literal truth of the Biblical account of Creation; and for a believer in a theology that held that humans were the pinnacle of God's creation, the choice of the name "Primates" was a natural one.

Linnaeus, however, was too good a biologist not to realize that humans, far from being separate from the rest of nature, form part of an intricate pattern that ultimately links all living things. Yes, like every other species, humans are unique. But humans also have much in common with other life forms. Linnaeus's solution to the problem of classifying life was to adopt a hierarchical approach that reflected the widening pattern of resemblances that he saw in nature—a pattern in which humans, for all their remarkable attributes, simply form just one more part. It was this consistency of pattern that allowed Linnaeus to classify those living species known to him in the same genus as those others that they most closely resembled; and in turn to group those genera into orders (and those orders into classes) according to their degrees of likeness to one another.

This system of inclusive grouping is reflected in Linnaeus's innovation of naming species by a pair of names, as in *Homo sapiens*. *Homo* is the genus name, and is unique; *sapiens* is the species name, and, in principle at least, is unique only in combination with the genus name. I'm not aware of any other species called *sapiens*, but there are, for example, any number of species whose second name is *africanus*: *Australopithecus africanus*, *Proconsul africanus*, *Kenyapithecus africanus*, and so forth.

In any event, the order Primates, like the genus *Homo* itself, was nothing special within the Linnaean system. It was simply the first-listed among the eight orders that Linnaeus recognized in his class Mammalia; and on Linnaeus's reckoning (which has, of course, been greatly refined and enlarged over the course of the last two-and-a-half centuries), it contained three other genera in addition to our own. His genus *Simia* contained the only apes then known, plus a gaggle of other species that we would refer to today as monkeys.

Somewhat less similar to ourselves are the members of his genus *Lemur*. For Linnaeus these included not just the primates of the island of Madagascar that we know as lemurs today, but also their fellow "lower primates" that live in Africa (the bushbabies and pottos) and Asia (the lorises), plus the much more distantly related colugos, gliding mammals from southeastern Asia. These latter are still often known as "flying lemurs" although they are not lemurs and do not fly.

The fourth genus that Linnaeus included within Primates was his *Vespertilio*, the bats. This inclusion of colugos and bats with the primates may seem odd, but is in fact coming to look increasingly prescient as more scientists reach the conclusion that the primates, colugos and bats form part of a single major subgroup within the class Mammalia.

The great beauty of Linnaeus's system of classification is that although it was conceived long before the advent of evolutionary thought, it fits perfectly with it. Linnaeus, an acute observer, derived his approach to classification from the pattern he saw in nature; what we have learned since is that this pattern is the product of the evolutionary process. Evolution, as neatly defined by Charles Darwin, is "descent with modification";

and the "nested" way in which our anatomy places us within a set of increasingly inclusive groups reflects both descent and modification.

Our closest primate relatives are the great apes, with whom we share a vast number of similarities; indeed, by one estimate we share some 99 percent of our genes with chimpanzees. This closeness of both anatomy and descent has led to our being classified with the great apes in the family Hominidae. The "lesser" apes, with whom we share a remoter common ancestry and fewer resemblances, join the hominids in the superfamily Hominoidea. In turn, we hominoids are joined by the monkeys of the Old and New Worlds in a larger grouping called the suborder Anthropoidea, and so on.

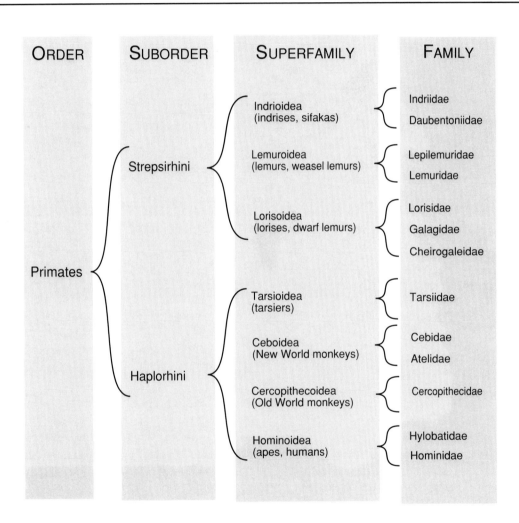

Classification of the Primates. Since Carolus Linnaeus first classified the order Primates into four genera in 1758, some two hundred living primate species have come to be distinguished. Not all primates are equally closely related, of course, and scientists continue to debate exactly how these two hundred species should be organized into larger groups. The simplified classification above represents as near to a consensus as one is likely to get. Note how the Linnaean system of classification creates an "inclusive hierarchy," in which lower ranks are clustered into higher ones (genera are grouped into subfamilies, which are grouped into families, and so on). Note also how the endings of superfamily ("-oidea") and family ("-idae") names help you recognize the ranking of the groups concerned.

A mongoose lemur, one of the "lower" primates from Madagascar, as drawn by the English naturalist George Edwards in 1754. It was on this illustration that Carolus Linnaeus, the father of modern animal classification, based the species *Lemur mongoz* in 1766.

The practical business of classifying living things is always in a state of flux, partly because if evolution involves change, then ultimately much of the anatomical evidence of relationship between species descended from a given common ancestor is going to disappear, or at least be obscured. This naturally leads to considerable difficulties for systematists, the biologists whose job is to describe natural diversity and to interpret its history.

It must also be admitted that these scientists have not made their job easier by failing to agree on whether classifications should be based on the "descent" or on the "modification" of Darwin's dictum. Some wish to see classifications reflect strict genealogies; others stress the amount of overall similarity between species. Because evolutionary change is an erratic process and is not a simple function of time, and particularly because similar structures tend to evolve in different lineages, classifications based on these two criteria can differ enormously. Regrettably, this is not a difficulty that will be resolved any

"Darwin and Friends." In this painting the artist Stephen Nash has shown a broad sampling of living primates. The primate species *Homo sapiens* is represented by the aged Charles Darwin.
Courtesy of Stephen Nash/Conservation International, and the Department of Anthropology, University College London.

time soon. While any group of life forms has obviously had only one evolutionary history, providing a target at which all systematists have to aim, classifications are not similarly products of nature. Instead, they are artifacts of the human mind, and human minds, being what they are, will continue to debate the criteria on which classifications should be based. Small wonder, then, that two-and-a-half centuries after Carolus Linnaeus and one-and-a-half after Charles Darwin, we are still grappling with the problem of defining the major group to which we belong.

Defining the primates

The order Primates, with at least a couple of hundred living species, is so amazingly diverse that it is difficult to point to features that all modern members of the order (let alone extinct ones) uniquely share. What all primates have most fundamentally in common is their descent from a single ancestor; but the evidence for exactly what that common ancestor would have looked like has become obscured by subsequent evolution among its descendants. This is why many specialists have preferred to abandon the search for a "definition" of Primates based on universally shared features that only primates have. Instead, they have followed the advice of the distinguished primatologist Sir Wilfrid Le Gros Clark (1895-1970) and have focused their attention on a number of physical trends that have marked the evolution of the order.

Although in many cases human beings happen to represent the most extreme expression of such trends, our species cannot be regarded simply as the inevitable culmination of a set of long-term tendencies. Evolution just doesn't work this way. Evolution is an opportunistic process that is greatly affected by factors that have nothing to do with adaptation to specific environments, and any trends that we may perceive in the fossil record exist, of course, only in retrospect. Species are not actors in a scripted drama; they are not fulfilling any kind of destiny. At any one time their major problem is one of simple survival—or, with luck, expansion—in a complex and irregularly changing habitat.

Still, it is undeniable that over the past 60 million years or so the evolution of primates as a whole has witnessed a certain number of trends such as those identified by Le Gros Clark. Chief among these are the enlargement of the brain relative to body size; the development of grasping and manipulative hands; the dominance of sight over smell; an emphasis on mobility in the design of the joints of the forelimb; the ability to hold the trunk of the body upright and to walk (or otherwise progress) on the hind limbs if necessary; and the development of complex forms of social organization that depend on elaborate behavioral signaling systems between individuals. These trends are readily discernible from the quick survey of primate evolutionary history in the next chapter.

What is a primate? Primates is the order to which humans belong, along with the greater and lesser apes, the monkeys of South America (the New World monkeys), the monkeys of Africa and Asia (Old World monkeys), the tiny, enigmatic tarsier of eastern Asia, the lorises and bushbabies of Asia and Africa, and the lemurs of Madagascar. The fossil record shows that other kinds of primates existed in the past, as well. Together with these, the two-hundred or so species of primates living today form a very varied assortment; but all belong to the same group because all share a single common ancestor, which probably lived in the Age of Dinosaurs, over sixty-five million years ago. In the long period since then many different adaptations have arisen among the primates. Some eat mostly fruit, others leaves, yet others gums and insects. Some move on four limbs, others move around mostly using their hind limbs, or their arms. Some prefer the trees, others the ground. Some live solitary existences, others live in pairs, or in small groups, or in groups numbering over a hundred. Some depend heavily on their sense of smell, others are more visual. Some weigh as little as two ounces, others reach four hundred pounds. Some are highly intelligent, others less so. But all are united by their common descent from that unique ancestor, back in the Age of Dinosaurs.

Chapter Four

Primate Evolution

Setting the stage

The disappearance of individual species is something that goes on, at a low level, all the time. But at occasional long intervals the history of life on Earth has been marked by large-scale extinctions, in which substantial proportions of all species living have suddenly disappeared from the geological record.

There has been active debate over the last decade or so over the reasons for such mass extinctions. Some scientists have seen the driving force of such events in the collision with Earth of large extraterrestrial objects, such as asteroids. Huge impacts of this kind, these scientists propose, caused so much dust and rubble to be heaved up into the Earth's upper atmosphere that incoming solar energy was largely blocked. Such energy fuels the plants and phytoplankton which form the base of the food chain, and these in turn are essential to all the other life forms that feed on them. Therefore, it is argued, any sudden deprivation of solar energy would have led to rapid ecological collapse, with widespread extinction of species. Explanations of this kind, which see major extinctions as geologically instantaneous happenings that occur over weeks, or years at the most, are most vigorously supported by those who detect a regularity in their occurrence (at approximately 26-million-year intervals).

However, as far as can be told from rather spotty evidence, the most consistent association of major extinctions is with global climate change, notably episodes of cooling, rather than with large impact events—though the latter may well have played a role in many episodes of extinction. For example, the dinosaurs had been declining in numbers and variety—and the mean temperature had been falling—for some millions of years before what the geological evi-

dence suggests was some kind of extraterrestrial impact marked their final demise at the end of the Cretaceous period, about 65 million years ago.

Whatever the case, the misfortune of the dinosaurs was a blessing to the mammals of the time, who found an immense variety of ecological opportunities opened up to them by the disappearance of the competition. Mammals had, in fact, been around for about as long as the dinosaurs; but while over many tens of millions of years the dinosaurs and their relatives had diversified to embrace a vast range of different lifeways, the early mammals hadn't really done much at all beyond survive. As far as we are able to judge from a rather poor fossil record, mammals diversified modestly at best during the Age of Dinosaurs that covered their first 130 million years of existence. For all of this time they largely remained small-bodied, rather shrewlike, probably nocturnal, and not particularly abundant.

All that changed with a vengeance as soon as the mammals had the stage to themselves, and the primates were among the first mammals to grasp their new opportunities. In this they were probably helped considerably by the fact that the last part of the Age of Dinosaurs also witnessed the spread of the flowering plants, with all of the new potential food resources that this implied. It's difficult to be certain about the origin of the primate order in the virtual absence of a fossil record from Africa, where, quite plausibly, it may have all begun at some considerable time before the dinosaurs died out. But the well-known fossil record from Europe and western North America bears—if anything at all—only the merest hints of early primates from the end of the Age of Dinosaurs. This is in contrast to a substantial abundance of them from the Paleocene epoch, the earliest part of the Age of Mammals, between about 65 million and 58 million years ago.

Our earliest ancestors?

These Paleocene primates, often known as plesiadapiforms (for their best-known genus, *Plesiadapis*), were a far cry not only from the apes and monkeys, but also from the lemurs, today's so-called "lower" primates which rely much more on smell than we and our closer relatives do, and have smaller brains and more-limited manipulative capacities. In fact, some scientists suggest that the "archaic primates" of the Paleocene are not specifically related to us, but instead simply form part of a larger group containing primates, bats, and colugos. Nonetheless, however the question may eventually be resolved, these early mammals probably give us a pretty good idea of what our own ancestors of that remote time were like.

All plesiadapiforms possessed large front teeth not found in later primates, but their chewing teeth were clearly of primate kind. The ancestral mammals had molar teeth with high crowns and tall, pointed cusps that were well suited for cracking insects open. The plesiadapiforms, by contrast, show some lowering and rounding of the cusps, suggesting that some at least had turned away from a primitive diet of insects toward a more general one that may have included a good deal of fruit. Some scientists argue that by the end of the Paleocene a range of food-processing adaptations was developing among the plesiadapiforms, some living mostly on fruit, others more on leaves, and yet others specializing on gums, sap, and the like.

In his mural painting in the American Museum of Natural History, the artist Jay Matternes has reconstructed several different primates from throughout the Age of Mammals. Here we see his re-creation of a species of *Plesiadapis*, among the best known of the "archaic" primates that flourished during the Paleocene epoch, between about 65 and 56 million years ago.

© 1993, Jay H. Matternes; original painting in American Museum of Natural History.

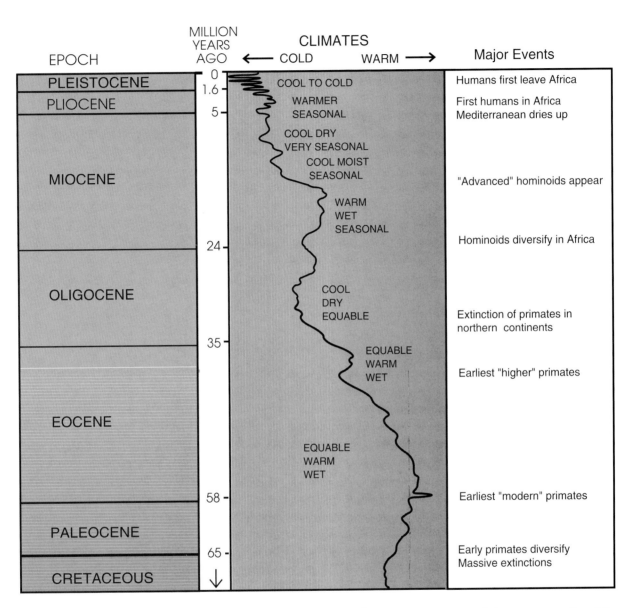

EPOCH	MILLION YEARS AGO	CLIMATES ← COLD WARM →	Major Events
PLEISTOCENE	0 1.6	COOL TO COLD	Humans first leave Africa
PLIOCENE	5	WARMER SEASONAL	First humans in Africa Mediterranean dries up
MIOCENE		COOL DRY VERY SEASONAL COOL MOIST SEASONAL WARM WET SEASONAL	"Advanced" hominoids appear
	24		Hominoids diversify in Africa
OLIGOCENE		COOL DRY EQUABLE	Extinction of primates in northern continents
	35	EQUABLE WARM WET	Earliest "higher" primates
EOCENE		EQUABLE WARM WET	
	58		Earliest "modern" primates
PALEOCENE	65		Early primates diversify Massive extinctions
CRETACEOUS	↓		

Geological Time. The dinosaurs died out about 65 million years ago, at the end of the Mesozoic ("middle life") era. The known record of primate evolution comes virtually entirely from the following Cenozoic ("recent life") era, which still continues. The Cenozoic is subdivided into a series of epochs, shown on this chart. Also shown is the climatic trend within the Cenozoic, which was a period of general cooling that eventually culminated in the "Ice Ages" of the Pleistocene epoch. The warmest times are shown in red, the coldest in blue. The wavy vertical line shows how temperatures have oscillated throughout the Cenozoic: the trend from warm to cool over this period was not a simple one. The world temperature curve is extrapolated from the ratio of heavy and light oxygen isotopes in the fossils of sea-dwelling organisms. During colder periods the percentage of the heavy isotope increases; it diminishes as the climate warms up. Major events in primate evolution are noted along the time scale.

Paleontologists disagree over the preferred environments of the various plesiadapiforms. Most likely, however, they lived both in the trees and on the forest floor, rather as squirrels do today. Robustly built, with relatively heavy bones bearing strong muscle scars, plesiadapiforms differ markedly from later primates in having had claws at the end of their digits. These suggest a mode of climbing very different from that of primates with grasping hands. In the skull, differences between the groups are even more striking. Among the plesiadapiforms the primate trend toward increase in brain size compared to that of the body had barely begun. The braincase was very small and was dwarfed by the face, which was long and capacious, suggesting a substantial reliance on the sense of smell.

In contrast, the eyes faced less completely forward than in later primates. This means that the fields of vision of the two eyes overlapped less, and such overlap is essential for stereoscopic "depth" vision. Neither were the eyes ringed by a circle of bone, a characteristic of all later primates. These features point to the retention of a relatively heavy reliance on the sense of smell, while in contrast the visual sense remained rather undeveloped. Whether some or all of the plesiadapiforms remained nocturnal, as their own ancestors certainly were, remains debatable, as does the kind of social organization that these mammals possessed; presumably some species were more gregarious than others.

One of the major problems we have in guessing about behavioral features such as this is, of course, that we lack totally any potential analogues among living primates. The same thing is, incidentally, true of the Paleocene mammals in general; the Paleocene was, as you might expect, a time of tremendous evolutionary experimentation among the newly expanding mammal order. During this time many new groups evolved which failed to survive far into the following Eocene epoch, when a variety of more familiar mammal types turn up for the first time. It's been suggested, for instance, that having achieved their widest diversification in the Paleocene, the plesiadapiforms were rapidly outcompeted in the Eocene by the newly evolved rodents rather than by their presumed cousins, the "primates of modern aspect."

The first "modern" primates

The Eocene epoch (58–35 million years ago) witnessed the disappearance of the plesiadapiforms and the rise of the first primates that resemble any living today. The beginning of the Eocene was probably the warmest part of the Age of Mammals, and during this time floras resembling today's tropical rain forest stretched well into regions that are temperate now. This effect was compounded by the fact that the geography of the continents at that time was very different from today's, many regions lying closer to the equator than they do now. In these vast humid forests there rapidly arose a huge abundance of primates of modern aspect, or euprimates ("true primates"), which quite closely resembled today's lemurs of Madagascar and the lorises and bushbabies of Asia and Africa. It is not clear who the ancestors of these new primates were, for known plesiadapiforms were too specialized to have played that role themselves, and there are no other likely candidates. Quite possibly the Paleocene primate fossil record of Africa, when it becomes known, will prove to hold the answer to this question.

These first euprimates are usually divided into two large groups, one believed to be related in a general fashion to today's lemurs and lorises, and the other to the enigmatic tarsier, a tiny-bodied and highly specialized nocturnal primate that survives today in parts of eastern Asia. This simple division almost certainly hides what is in fact a highly complex situation, but it will be a while before anything coherent comes along to replace it.

While in more traditional classifications of living primates the lemurs and tarsier are classified together as "prosimian" or "lower" primates, it has recently become quite common to find the tarsier ranged alongside the "higher" primates in the suborder Haplorhini. However, even if the tarsier does share an exclusive if remote common ancestry with the monkeys, apes, and ourselves, it is nonetheless clearly separate from that group. In any event, it is undisputed that the euprimates of the Eocene represent a radical change when compared to the plesiadapiforms. Thus, for the first time we find primates with opposable thumbs and great toes, and whose digits show clear evidence of having been tipped with flat nails backing up sensitive fingertip pads. The grasping hand had arrived, representing a complete change in strategy for moving around in the trees when compared with the clawed plesiadapiforms.

While studies of plesiadapiform limb bones always elicit comparisons with rodents of various kinds, the skeletons of Eocene euprimates recall those of modern lower primates. Some North American forms, in particular, tend to have limb proportions reminiscent of the sifakas of Madagascar, with greatly elongated hindlimbs and evidence of much mobility in the forelimbs. Sifakas practice a spectacular form of locomotion in the trees known as "vertical clinging and leaping," in which they leap enormous distances between vertical tree trunks or oblique branches, propelled by their powerful hind limbs. In the outer branches of the trees, where they spend much time searching for food, they show tremendous acrobatic ability, suspending themselves in every conceivable position. On the ground they bound along on their long legs, holding their trunks upright.

The skulls of Eocene euprimates testify with equal eloquence to a marked change compared to the plesiadapiforms. Perhaps most strikingly, they show a dramatic shift in the importance of vision relative to smell. Whereas plesiadapiforms retained the ancient mammalian primary reliance on smell, the early euprimates were clearly much more visual and, almost certainly, most were active during the daytime. The braincase had increased in size relative to the nose and face, and the eyes themselves were encircled by a solid ring of bone.

The overall appearance of these primates was thus not too unlike that of the lemurs and other lower primates living today. In these latter, of course, the de-emphasis of smell relative to vision is not as far advanced as it is in the monkeys, apes, and ourselves. Lemurs examine objects with their noses as much as with their eyes, and they manipulate them much less dextrously than higher primates do. They also have rather smaller brains compared to body size. And while the lemurs manifest many different forms of complex social organization and behavior, communication by olfactory "marking" is important to them in a way that it is not to almost all higher primates. In contrast, higher

---

"Modern" Primates. Today's primates are very different from the earliest members of our group. But at some point over 55 million years ago primates appeared on the scene that were not very different from the "lower" primates of today: the lemurs of Madagascar and the lorises and bushbabies of Asia and Africa. Compared with the more archaic types, primates of this new kind had larger brains compared to their body sizes; smaller noses, indicating a less complete reliance on the sense of smell; eyes that were supported by a complete ring of bone and that faced forward, producing a good degree of stereoscopic "depth" vision. Perhaps most important of all, these early "true primates" boasted grasping hands, in which the thumb was moved away from the other fingers. Gone were the primitive claws, replaced by flat nails which backed touch-sensitive pads at the ends of the fingers. These linked innovations provided a new ability to exploit all of the different ecological zones offered by the tropical forests, and laid the basis for the refined manipulative abilities that were ultimately so important in the evolution of humans.

primates have more complicated visual signaling behaviors, often involving facial expression—with a specialized facial musculature to provide it.

Lemurs thus have in many (though far from all) ways remained rather conservative relative to the higher primates; and by judicious analogy they can tell us much about what our Eocene precursors were like in life. In particular, we can glean from them that the essential primate sociality—and the adaptability of social organization and behavior that goes along with it—was almost certainly well established early on in the Eocene. And it is in sociality of typically primate kind that we can glimpse the remote origin of our own complex human societies.

Early "higher" primates

Not one of the American or Eurasian Eocene primates currently known presents us with the combination of features that comparative anatomy suggests we should find in the ancestors of the Old and New World monkeys, apes, and humans. Actually, the search for characteristics that one might expect to occur in the common ancestor is not an easy one when we're looking at a group as diverse as the suborder Anthropoidea. This group unites the "platyrrhine" (or New World) monkeys of South America (themselves as different as marmosets and spider monkeys) with the "catarrhine" primates. These latter consist of the African and Asian (or Old World) monkeys (which range from Africa's baboons to Borneo's bulbous-nosed proboscis monkeys), the gibbons or "lesser" apes, the "great" apes (chimpanzees, gorillas, and orangutans), and ourselves. Their tongue-twisting names come from the contrasting nose shapes of the New and Old World forms, but the two differ in many other ways as well. Yet most primatologists have a hard time believing that the numerous shared similarities of their (otherwise different) ear regions could have

Painting by Jay Matternes depicting a group of early "primates of modern aspect" belonging to the genus *Notharctus*. This early primate was common in the rain forests that covered Wyoming during the Eocene epoch, some 48 million years ago. Its long hind limbs suggest that, like today's sifakas of Madagascar, it was an agile leaper between vertical supports.

© 1993, Jay H. Matternes; original painting in American Museum of Natural History.

evolved twice, for the ear is a notoriously complex structure. But to believe that such a structure was acquired by descent means that it was acquired from a common ancestor. Where—and when—did that ancestor live?

The "when" is the (possibly deceptively) easy part: despite the lack of fossils, most paleontologists look to some time in the late Eocene, perhaps around 40 million years ago or a little more. Unanimity rapidly disappears, however, when we ask where.

During the Eocene and pretty much throughout the Oligocene epoch (35–24 million years ago) which followed, the African continent was isolated by a sea barrier from Eurasia. In turn, an ocean existed between Eurasia and North America, and between North

A modern sifaka, *Propithecus verreauxi coquereli*, photographed in a deciduous forest at Ampijoroa, Madagascar. These beautiful creatures are not closely related to Eocene primates such as *Notharctus*, but move through the trees in a similar way.

Photo by Ian Tattersall.

America and South America. The descendants of any late Eocene ancestor must thus have crossed at least one major sea barrier to find themselves eventually in both the Old and New Worlds. But which sea?

This remains uncertain, although the smartest money is on an African origin followed by an ocean crossing to South America (from which, despite a pretty good fossil record, primates are not known before the late Oligocene). For in spite of the huge watery gap that looms when one looks at modern world maps, falling sea levels following the end of the Eocene may have exposed large parts of the mid-oceanic ridges that almost connect the two landmasses.

Clearly, though, our perceptions are greatly colored by the fact that our knowledge of Oligocene primates comes almost entirely from one area of the world: a tiny part of North Africa. For the end of the Eocene was marked by another of those periodic large extinction events, and primates from many parts of the world were caught up in it. Following the Eocene, primates promptly disappeared from Europe, and more gradually from North America; and from this point on Africa assumes an increasingly important role in our view of primate evolution.

What is a "higher" primate?

Before we look at this new African evidence, let's pause for a moment to consider some of those characteristics that do seem to unite today's higher primates. Firstly, all to one degree or another further accentuate the tendencies that are already evident in the Eocene euprimates, and which we see in the lemurs today. They share, to varying degrees, a significant enlargement and elaboration of the brain, and highly accurate color vision, with the eyes completely encased by bone at the sides and the rear of the orbit. There is almost complete overlap of the visual fields of the two eyes, and the sense of smell is yet further diminished, with reduction of the nose cavity and the loss of the "rhinarium" (the external "wet nose" that is familiar in dogs as well as retained by lemurs). The hands have (in most cases) an opposable thumb that permits more precise manipulation of objects than the "whole-hand" grasping typical of lemurs. And social systems are highly organized, with subtle and complex relations among their members. Only one higher primate—the owl monkey of South America—is nocturnal, and it probably acquired this habit secondarily, while actually evolving from a diurnal ancestor.

Ancient higher primates of the Egyptian forest

There is much debate over which of several fossils is the earliest that could be said to belong to a higher primate. However, all contenders other than those from Egypt's Fayum region are fragmentary bits of jaw and teeth, and are either unconvincing in themselves or don't really give us the evidence we need to be sure. Some of these fossils have been known for years to be of late Eocene age; but since it has become likely that some of the earliest primates from the Fayum may be from the Eocene too, their potential importance has paled still further in comparison to the Egyptian fossils.

The paleontological potential of the Fayum has been recognized since the latter part of the nineteenth century, and the first specimens representing new kinds of fossil primates were found there in the first quarter of this one. However, the true significance of this unique area did not begin to emerge until, in 1961, Elwyn Simons of Duke University (then of Yale) began intensive field work that continues to this day. Simons summarizes this significance succinctly: "The fossil primates of the Fayum," he says, "are so diverse that they suggest that the earliest radiation of higher primates, and perhaps of primates in general, took place in Africa. The Fayum is one of the most incredibly prolific fossil primate sites in the entire world." Besides amassing a vast treasure-trove of fossils, Simons and his colleagues have shown that all of the Fayum primates are at least 31 million years old, and that the most ancient of them are possibly as much as 35–40 million years old. The area to the west of the oasis known as the Fayum is today part of Egypt's arid Western Desert, and lies well over one hundred miles south of the shores of the Mediterranean.

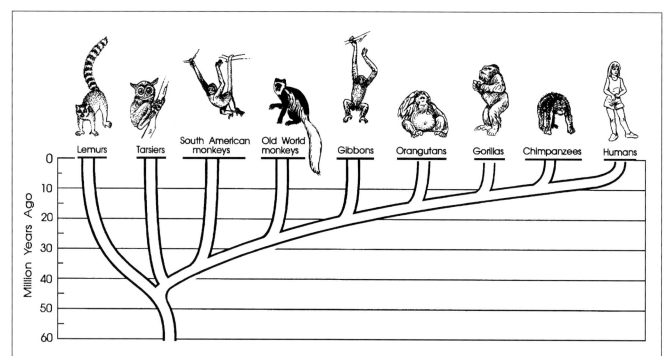

Diagram showing the approximate evolutionary relationships and times of divergence of today's major groups of primates. The main uncertainty in this scheme is the place of the tarsier; some claim that its closest affinities are with the lemurs and lorises, rather than with the "higher primates," as shown here. Many other details of primate evolution also remain unresolved. *Illustration by Diana Salles.*

Skull of the Eocene primate *Notharctus*, recently discovered in Wyoming by a team of paleontologists from the American Museum of Natural History discussed on pages 46–47.

Photo by Peter Goldberg; courtesy of John Alexander and Natural History.

But in the Oligocene it lay at the northern fringe of the African continent and was clothed in dense tropical forests through which broad rivers meandered. In these forests lived a variety of primates. Nearly all of them qualify to be considered as higher primates, although one appears to be broadly related to certain Eocene lower primates, and another more specifically to the tarsier. Most Fayum primate fossils, however, represent one of two families: Parapithecidae and Propliopithecidae.

The parapithecids were small-bodied by higher primate standards and, interestingly, show an ear structure and many other features that resemble those of the South American monkeys. Some paleontologists have thus suggested that the parapithecids may be related to the New World monkeys, but for a variety of reasons most would consider that they represent an early side-branch of higher primate evolution without any special relationship with any of today's higher primates.

The best-known propliopithecid is *Aegyptopithecus*. This primate was not described until 1965, but thanks to the labors of Professor Simons and his colleagues it is now known from several dozen fossils. The size of a big house cat, *Aegyptopithecus* had a small brain not much, if at all, larger relative to its body size than those of lemurs. But its snout, though on the large side, apparently lacked the elaborate sensory apparatus typical of modern lower primates, and the area of its brain devoted to smell was reduced. Small eye sockets, fully enclosed at the rear, hint that it was active during the daytime.

Among living primates, the skeleton of *Aegyptopithecus* most closely resembled South America's howler monkeys, and analysis of its teeth suggests that it mostly ate fruit. Males were bigger than females. This is consistent with a complex social structure in which groups contained numerous adult males and females, the males competing for females rather than pair-bonding. Although the propliopithecids had a New World monkey-style ear region, as in the case of the parapithecids this does not appear to indicate any special relationship with the South American primates: rather, this kind of ear structure was the original higher primate condition, from which later variants

Geographical relationships between Africa and South America at approximately 40 million years ago. At that time the two continents were much closer to each other than today, while North and South America were separated by a wide expanse of ocean. Ocean ridges are possible sites of island chains. Arrows show directions of ocean currents.

Illustration by Diana Salles after a reconstruction by D.H. Tarling.

evolved. Thus, taking all the features of the skull and skeleton into account, most paleontologists would nowadays consider that *Aegyptopithecus* and its kin were equally related to both Old and New World living primates—if, indeed, they were not close to the common ancestry of both.

Hominoids at last

The Miocene epoch (24–5 million years ago) witnessed many important geological events, not least among them the development of the great East African Rift System, a huge complex tear in the earth's surface that stretches from the Red Sea in the north to Mozambique in the south. New continents are created and move away from each other as long upwellings of molten material from deep in the earth crack the land surface and force the resulting fragments apart. As the precursor of the yet-to-occur separation of the eastern and western parts of the African continent, the East African Rift is a classic example of this process, a process which, for paleontologists, has two enormous advantages. First, the heaving-up of the landscape that it entails increases erosion and thus the supply of the materials that make up the sedimentary rocks in which fossils are found. And second, rifting is accompanied by a great deal of volcanic activity, which introduces layers of volcanic rock into the accumulating sediments. Unlike sediments, such volcanic rocks can be directly dated using a variety of modern techniques—of which there will be more discussion later.

From the early Miocene on, right up to quite recent times, sedimentary rocks associated with the evolution of the East African Rift play an incomparable role in providing us with fossil evidence of the evolution of humans and their forebears. During the first part of the Miocene, Kenya and Uganda were relatively low-lying areas covered by humid forests. These forests supported a diversifying fauna of primates that included the earliest forms that we can classify, along with the greater and lesser apes and ourselves, in the superfamily Hominoidea. Thus sites in Kenya and Uganda associated with rift volcanism,

Group of early Old World higher primates of the genus *Aegyptopithecus*, well known from the Fayum region of Egypt. Approximately 34 million years old, the remains of this primate are found in sediments which also yield abundant fossilized fruits of the vine *Epipremnum*, on which *Aegyptopithecus* is seen feeding here.

© 1993, Jay H. Matternes; original painting in the American Museum of Natural History.

Some of the best-known fossils of *Aegypto-pithecus* found in the Fayum of Egypt. The discovery of a remarkable series of fossil faces of *Aegyptopithecus* showed that the first-found skull (second from left) had an unusually long face (probably due to distortion during fossilization), and that males were more robust than females, with larger canine teeth.

Photo courtesy of Elwyn Simons.

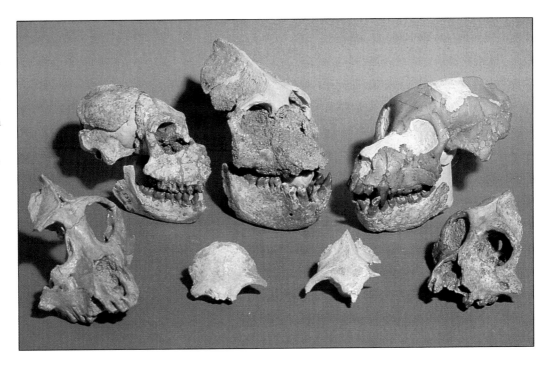

and dating to about 23–16 million years ago, have produced a remarkable series of fossils in which primates of the hominoid family called the Proconsulidae were an important element. Indeed, it was during the early and middle Miocene that the hominoids achieved their greatest diversity, with a much greater variety of species than survives today.

The living apes

To put the early hominoids in context, we should make a few observations about the surviving apes which, though relatively few in species, make up a very varied group. The nine species of gibbons, the lesser apes, are relatively small-bodied. Their weight varies from twelve to twenty-four pounds. They live in the moist rain forests of southeastern Asia and subsist principally on fruit, supplemented with young leaves and various invertebrates such as termites. Without doubt, their most striking characteristics are their greatly elongated slender forelimbs and long, hooked hands and feet; these are specializations for the spectacular arm-swinging mode of locomotion known as brachiation. Males and females are of similar body size, and both have long, sharp canine teeth; this lack of difference between the sexes is apparently related to a social organization in which adult males and females form permanent bonds. The young remain in the parental group only until they mature. Gibbons are classified in their own family, Hylobatidae, distinct from great apes and humans.

Also Asian is the much larger orangutan, of which the single surviving species is restricted nowadays to rain forests on the islands of Sumatra and Borneo. Unlike the gibbons, the orangutan displays a striking size difference between the sexes, females weighing about 130 pounds and males twice as much or even more. Orangutans are highly specialized for relatively cautious movement in the trees, using all four limbs. They have long forelimbs, short legs, and long hands with curved fingers and a short thumb. Primarily fruit-eaters, they round out their diet with a wide variety of other kinds of tree produce. Larger males, especially, spend a lot of time on the ground, where they move around with their hands curled up into a fist.

Unlike other hominoids, adult orangutans lead rather solitary existences, the only consistent social grouping being that of the adult female with her immature young. Individual females live in relatively small home ranges of about 180 acres, and tend not to move around a lot; the ranges of adult males are larger, overlapping those of several females. Young adult males roam around until they acquire their own territory. The age at which they mature fully depends on social factors such as when they acquire territories. Sexual maturity occurs between the ages of eight and fifteen years. Females mature slightly younger. This long period of maturation and of dependence on the mother reflects the expansion of the juvenile learning period common to all the great apes. This is related in turn to the complex social contexts in which hominoids must learn to live and to the sophisticated exploitation of the environment that is their hallmark.

Chimpanzees are found in Central African environments that range from tree savanna to rain forest. They show much less size difference between the sexes than do most other great apes (male adults weigh in at about 100-120 pounds, females at 70-90 pounds), and the fore and hind limbs are more nearly equal in length. As much at home in more open woodland areas as in closed forests, chimpanzees tend to travel on the ground where, like gorillas, they "knuckle-walk," bearing the weight of the front of the body on the first knuckles of the closed hand. Fruit is their primary food, but chimpanzees are opportunistic feeders and even prey on other mammals, including other primates. Individual chimpanzee adults tend to spend a good deal of time foraging on their own, and join up with other individuals to make up temporary groupings of widely varying composition. The bonobos, close relatives of the chimpanzees, also show relatively small body size differences between the sexes but are strictly rain forest denizens.

Gorillas, which are confined nowadays to small areas of humid forest in Central Africa, exist in better-defined social groupings of about a dozen individuals. Each group usually contains one mature male, one or two younger males, and several adult females with their offspring. Adult males may weigh 350-

A gibbon (above), the smallest of the hominoid primates, and the most distantly related to humans. Gibbons are most strikingly distinguished by their long arms, related to their highly specialized "brachiating" form of locomotion, suspended under branches and swinging arm over arm.

Illustration by Diana Salles.

A female orangutan (right). Females of this ape are much smaller than males and move with greater agility in the trees.

Illustration by Diana Salles.

A male chimpanzee holding a twig, such as those used in "fishing" for termites.

Illustration by Diana Salles.

pounds and up, and females 200 pounds plus. Predictably for such large primates, gorillas are quite terrestrial: adult males, in fact, spend nearly all of their time on the ground. Gorillas eat large quantities of herbaceous vegetation; and indeed many of their famous displays that involve tearing up plants or breaking branches have the effect of encouraging new growth to shoot up—a simple form of environmental control!

Variety, then, is hardly lacking in the diet, social organization, and other features of modern hominoid life. This is particularly remarkable when we realize how few species are involved, for only one living hominoid species—*Homo sapiens*—has escaped mention above. But if variety is judged by number of species, Hominoidea has fallen on very hard times. In contrast, the Miocene, the very first part of hominoid history, was the time when our superfamily achieved its greatest abundance of species. This occurred well before the diversification of the Old World monkeys got under way, the latter coinciding with a sharp decline in number of hominoid species as environments changed toward the end of the Miocene. In any event, several different kinds of hominoids of the family Proconsulidae have been distinguished among fossil primates of the early to middle Miocene. By far the best known among these is the eponymous genus *Proconsul*. A. Tindell Hopwood, who first described this genus in 1933, thought it ancestral to the chimpanzee and named it for Consul, a circus chimpanzee famous throughout pre-war London for his smoking and drinking habits.

A male mountain gorilla from Rwanda. Note the knuckle-walking posture with the weight of the foreparts borne on the first knuckles of the closed hands.

Illustration by Diana Salles.

Proconsul and its relatives

View by Jay H. Matternes of the Miocene forest near Rusinga Island, Kenya, showing a species of the rather primitive hominoid genus *Proconsul*, which flourished there about 18 million years ago. See illustration, page 57, for an explanation of how these primates were reconstructed.

© 1993, Jay H. Matternes, original painting in the American Museum of Natural History.

1948 the front part of the skull of a member of the smallest *Proconsul* species was found by Mary Leakey at a site on Rusinga Island, a speck of land close to the Kenyan shore of the world's second-largest lake, Victoria. Fossils from this site date from about 18 million years ago. In the previous year some other skull bones had been collected at the same place, but they lay unrecognized in a museum drawer until the middle 1980s, when the paleontologist Martin Pickford realized that they were from the same individual as the 1948 skull. Once all the pieces were glued together, thirty-five years after the original find, the paleoanthropologist Alan Walker and two colleagues were able to conclude that this primate had a brain of about 170 milliliters in volume. Relating this figure to an estimate of body size derived from some limb bones that had been recovered at Rusinga in 1951, these researchers concluded that *Proconsul* had had a brain larger than that of modern monkeys of comparable size.

The interest which these museum-drawer "discoveries" excited led to further field-work at Rusinga in 1984, and this resulted in the finding of yet more pieces of the same skeleton, along with a trove of other *Proconsul* fossils at a nearby site. The upshot is a richness of documentation of this *Proconsul* species that is altogether remarkable in the primate fossil record.

Preliminary analysis of these various finds revealed an unexpected picture. While early researchers such as Hopwood had believed that *Proconsul* and its relatives were in some way ancestral to the living African apes, the new fossils suggested a different interpretation. For although *Proconsul* resembles the chimpanzee in a few features of the body skeleton and dentition, it surprisingly shares many more skeletal resemblances with

Reconstructing an extinct primate. Reconstructing the appearance of extinct primates from their fossil bones is a complex process that requires a lot of specialized anatomical knowledge as well as artistic skills. Nearby appears the reconstruction of the Miocene primate *Proconsul* made by artist Jay H. Matternes for his mural at the American Museum of Natural History. Here we glimpse some of the background work that made that final reconstruction possible. First Matternes prepared a complete drawing of the skeleton. This had to be pieced together from an incomplete fossil skeleton, supplemented with information from other individuals and filled out from experience based on many dissections. Next, the musculature was carefully reconstructed over the bony framework, based on the artist's extensive comparative anatomical knowledge of living primates. Only then was the external appearance determined, from the bulk of the muscles and other tissues underlying the skin. For the final painting, which shows the living *Proconsul* in action, such essential aspects as posture and facial expression had to be determined from studies of how the muscles could have moved the joints in the living individual, and from knowledge of the facial musculature in related living primates. With all of this scientific rigor, however, some judgments remained subjective. Such things as fur color and ear shape cannot be determined from the evidence of the bones, and they were supplied on the basis of what "felt" right to a highly experienced artist and anatomist.

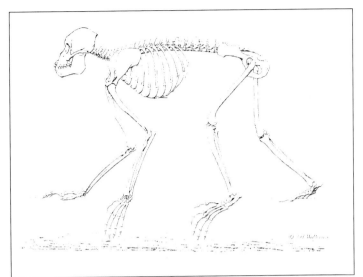

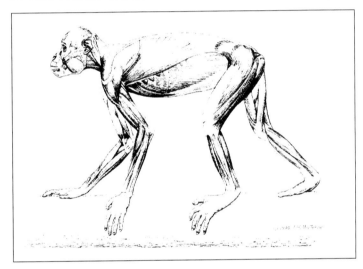

the Old World monkeys. These similarities most likely represent a set of early Old World higher primate features that have been lost more thoroughly by modern hominoids than by modern Old World monkeys. Like the living hominoids, *Proconsul* lacked a tail, but at the same time it also lacked most of the specializations we see in modern great and lesser apes.

A picture is thus emerging of these early hominoids as rather unspecialized arboreal quadrupeds: robust and relatively slow-moving primates that were less agile than living hominoids, and which showed none of the special adaptations for suspension in the trees or for moving on the ground that the great apes possess. There is, for example, no evidence in their skeletons for knuckle-walking, even though the larger species may have been partly terrestrial. Yet in terms of descent, the affinities of the proconsulids are clearly with apes rather than with monkeys. The proconsulids, then, retained an ancestral Old World higher primate anatomy that was related to a preference for fruit-eating in moist forest habitats. Later hominoids subsequently found more specialized methods of surviving within this (shrinking) ancestral habitat, while in contrast the specializations of the Old World monkeys may have been acquired as a result of shifting into more open-country environments.

In any event, many scientists today believe that *Proconsul*, rather than being an ancestral African ape, is equally related to all of the modern apes: the gibbons as well as the great apes and humans. This theory fits with the idea that certain other Miocene hominoids previously thought to be ancestral gibbons are in fact unrelated to the living lesser apes: once again, resemblances that appeared to point that way are now thought to be common holdovers from a remote ancestor, not features inherited from a specific common ancestor. Evidently, the lineages leading to the great and lesser apes, let alone to humans, separated after the time of *Proconsul*—which must therefore occupy a position at least reasonably close to our own remote ancestry.

One could wish (as always!) for a better fossil record; but let's look next at the rather poor evidence that does exist for the ancestry of the great apes.

Early great apes

Until recently it was taken for granted that humans (along with their close fossil relatives) form a single coherent unit that contrasts with another group formed by the great apes, our closest living relatives. Largely as a result of two factors, all this has changed in recent years. One of those factors has been the introduction into studies of evolutionary relationships methods that compare genes instead of anatomical characters. Most (though not quite all) analyses of molecules and of DNA sequences have suggested that the African apes are more similar to us in the systems tested than we—or they—are to the orangutan. If this is so, the African apes and ourselves should be placed in a single group distinct from the orangutan.

This new approach more or less coincided with the second new factor, the arrival in paleoanthropology of cladism, a method of reconstructing evolutionary histories that emphasizes the importance of descent and common ancestry rather than simple overall similarity. The word "cladism" comes from the ancient Greek word for "branch"— cladists love to draw branching diagrams to show how organisms are related. Looking at the hominoids from a cladistic perspective reveals a picture very different from what we see when we look at the great apes and ourselves purely in terms of general resemblance. It's quite true, of course, that at first glance the great apes all seem to resemble each other more than any of them does us. But a moment's thought shows us that most of the differences separating us from them are unique properties of our own lineage: that, in other words, we have changed more since our shared common ancestor than has any of the apes. To put it the other way around, the three great ape genera simply resemble each other because all remain more similar to our common ancestor than we do, and not because of any special relationship that exists among all three.

Since eliminating both ancestral and unique characteristics from consideration leaves us with a greatly reduced quantity of evidence on which to base our theories of hominoid interrelationships, it's almost inevitable that the details of these relationships have been endlessly debated.

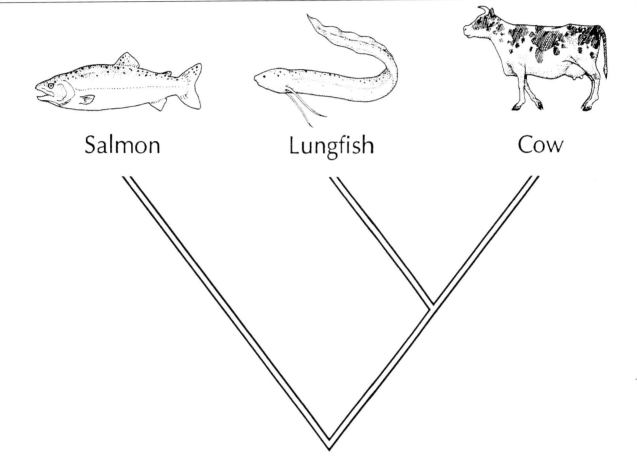

Cladistics: A Way of Determining Evolutionary Relationships. Traditionally, determining the evolutionary relationships among a group of species was an intuitive affair largely involving overall similarity of appearance. Cladists point out that this can be misleading. The classic example is to take three vertebrates: salmon, lungfish, and cow. Which pair is most closely related? If you voted for the salmon and the lungfish, you're wrong. Lungfish and cows actually share a common ancestor more recently than either does with the salmon and are thus more closely related. This is expressed in the branching "cladogram" above. The superficial similarity between the aquatic lungfish and salmon simply means that both have departed less from the shape of the aquatic common ancestor of all three than the terrestrial cow has. Cladists thus insist that in figuring out evolutionary relationships we should look only at those characteristics that are due to descent from a unique common ancestor. For example, while we can say that all vertebrates are descended from a common ancestor with a backbone because all possess a spine, we have to acknowledge that within Vertebrata having a backbone tells us nothing about relationships: if we want to know how different vertebrates are related we have to abandon this feature and look at other ones.

The late Miocene hominoid *Sivapithecus*, well known from the sediments of the Siwalik Hills in Pakistan and India between 12 and 7 million years ago. This *Sivapithecus* species quite closely resembles the orangutan in the structure of its jaws and face, but its body skeleton remains quite unspecialized, with rather monkey-like proportions. Unlike the earlier hominoids, which were adapted to life in dense forests, *Sivapithecus* flourished in a relatively open woodland environment.

© 1993, Jay H. Matternes, original painting in the American Museum of Natural History.

Jeffrey Schwartz of the University of Pittsburgh, for example, created quite a furor by suggesting that our closest relative in terms of descent is the orangutan (which any zookeeper will confess is the most ingenious of the apes). "When you look beyond the conventional characteristics usually considered," says Schwartz, "you find that orangs share more features with humans than do any of the other apes. Particularly intriguing commonalities include the lack of estrus [cyclical sexual receptivity] in the female, the long gestation period, marked asymmetries between the two sides of the brain, and very high levels in pregnant females of the hormone estriol, an index of proper brain growth in the fetus."

Provocative though Schwartz's views are, though, it's more commonly thought that chimpanzees are our closest relatives, a notion that received a considerable boost when the geneticists figured out that we share more than 99 percent of our genes with them (though the fact that we share about 99.9999997 percent of our total evolutionary history with them may help to put this striking number in perspective). Yet most scientists probably still prefer the notion that humans are descended from a common ancestor that only later gave rise to the ancestor of the chimpanzee and gorilla.

Unfortunately, the fossil record doesn't help much in resolving this problem. In East Africa, at about 14 million years ago, the fossil mammals we find change dramatically. This new fauna is more like the one familiar today and suggests an environmental change to drier and more open woodland habitats. Among the hominoids, the primitive *Proconsul*-like types yielded to species whose teeth, in particular, indicate a shift from eating fruit to diets of the tougher, more resistant foodstuffs available in the new environment. Toward the end of the middle Miocene the African and Eurasian continents made contact with each other, and interchange of animals between the two landmasses became possible. As a result of this, hominoids of this new kind are actually best-known not from Africa but from Eurasian sites dating in the 14–7-million-year range. Indeed, a late survivor, *Gigantopithecus blacki*, is known from Chinese sites that are under a million years old.

As its name hints, this is almost certainly the largest primate that ever existed: extrapolation from the few massive jaws known suggests to the anthropologist Russell Ciochon and his colleagues, who made a life-size construction of the beast, that *Gigantopithecus* males may have stood over ten feet tall and weighed over twelve hundred pounds! Remarkably, this extraordinary creature was first identified from teeth obtained from traditional Chinese drug stores, where they were sold as "dragon bones" and prized for their medicinal effects when ground to powder.

Of the later Miocene hominoids, the genus *Sivapithecus* ("Siva's ape," named for the Hindu god of destruction) is without doubt the best documented by fossils. It is known from sites in both Africa and Asia, but most particularly from fossil beds in the Siwalik Hills of India and Pakistan. These are foothills of the Himalayas, the great mountain range which began building in the middle Miocene as India collided with Eurasia. It is findings in Pakistan in the 1970s and 1980s that most clearly led to the recognition of *Sivapithecus* as a relative of the orangutan, although this relationship had been hinted at ever since the first discoveries of *Sivapithecus* in the late nineteenth century.

A comparison of an approximately 8-million-year-old face of a male *Sivapithecus indicus* with that of a modern orangutan instantly shows the numerous similarities in this complex region between these species. The tall and closely spaced eye sockets and the dished-out lower face are novel characteristics which even by themselves distinguish these two species from the African apes and plainly suggest common descent. However, it is striking that *Sivapithecus indicus* did not share the remarkable specializations of the body skeleton that characterize the modern orangutan. Instead, the body is that of a rather generalized great ape, more similar to the African apes than to the orangutan (presumably because of retained ancestral characteristics), but without any of the specializations that the African apes show for knuckle-walking.

As for the African apes, no clues from this period exist about their origins. None of the middle to late Miocene hominoid fossils known from East Africa offers itself as a con-

Fossil face of *Sivapithecus* from the Siwalik Hills of Pakistan dated to about 8 million years ago. It is remarkably similar in many details to the recent orangutan skull shown on the right.

Photo by Willard Whitson.

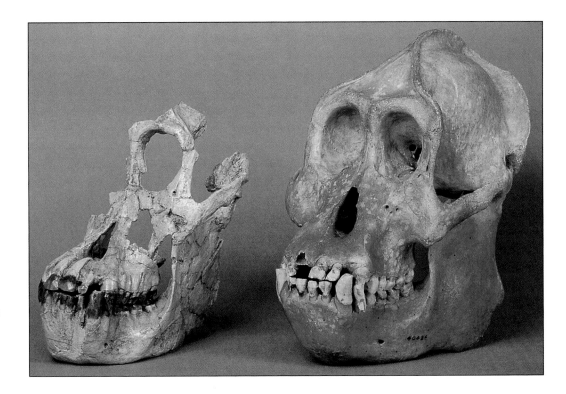

The sites where early
hominoid fossils have
been discovered.

*Illustration by
Diana Salles.*

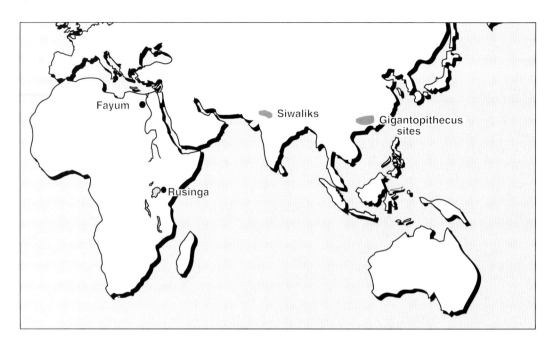

vincing candidate for their ancestry. All we can tell from the fossil record is that hominoids became less abundant as the Miocene wore on. At this time the area of the East
African Rift was doming up, forming highlands running from north to south. These
trapped moisture-laden winds from the west, and eastern Africa found itself increasingly
in their rain shadow. In this drying environment the distributions of the ancestral apes
presumably shrank along with their rain forest habitat. They appear to have left the
exploitation of the new, drier environments to the Old World monkeys, which became
more diverse and abundant as the hominoids themselves declined.

Nor do we have in this period any convincing evidence for early human relatives. At
one time *Ramapithecus* (another fossil named for a Hindu deity), which is known from
both Africa and Asia in the period from about 14–10 million years ago, was thought to be
a potential human ancestor. Today, though, *Ramapithecus* is believed to belong among
Sivapithecus and its relatives, some of which, at least, seem to have been tied in with
orangutan ancestry.

The last Eurasian hominoids (the *Gigantopithecus* lineage excepted) seem to have
died out by around 7 million years ago, leaving a gap in the fossil record of our larger
group until some time before about 4 million years ago. When that record picks up again
we find ourselves back in Africa, where the earliest fossils that identifiably lie within our
own lineage are found. And at this same point the focus in our examination of human origins subtly shifts, from looking at those things that humans have in common with other
denizens of the living world, to the history of those features that make us unique.

Chapter Five
Human Evolution

When we think of those extraordinary qualities of the mind that distinguish us so clearly even from our closest living relatives, we human beings find it tempting to see ourselves as the culmination of a long and gradual process of perfection. Even recent scientists have fallen into this trap. Back in the 1960s and early 1970s, for instance, many argued that the acquisition of culture made humans so broadly adaptable that not more than one human species could have existed on Earth at the same time. And though later fossil discoveries laid to rest any literal adherence to the idea that our evolutionary history consists of a simple succession of species, many scientists still try to envision the simplest picture possible, with an absolute minimum number of fossil human species. This reflects the durability of that old theme the *scala naturae* we mentioned earlier: human fossils, like our primate relatives, are seen simply as pausing-points along an almost inevitable road to perfection. Yet to look upon our evolution in this way is, of course, to misunderstand what that complex process is all about. Let's see why.

The young Charles Darwin, drawn by Diana Salles after a painting made just before Darwin left on his historic round-the-world voyage on the H.M.S. *Beagle*.

How evolution happens

The traditional view of evolution is to see it as the product of steady environmental pressure exerted on each species, improving its adaptation to its habitat. This pressure, called natural selection, works like this: In any population, no two individuals are exactly alike. Most of the differences between them are inherited from their parents, and are thus capable of being passed along to the next generation. What's more, in any generation more individuals are born than will survive to reproduce themselves, and this is true even of such rapidly expanding populations as that of *Homo sapiens*. Those who do reproduce most successfully are generally those best able to cope with the surroundings in which they find themselves; and the genes that confer such adaptive advantage will be preferentially represented in the next generation. Each generation will thus be slightly different from the one before it, a little better adapted to the prevailing conditions.

Perhaps the most famous example of this is the peppered moth, *Biston betularia*, that lives in the north of England. Normally this moth is pale-colored and is hard to see against the whitish lichens that cover the trees on which it rests. Its coloration thus confers protection from insect-eating birds. The occasional dark-colored moth, on the other hand, is spotted instantly by birds and weeded out of the population, removing the gene responsible for its coloration. However, from the beginning of the Industrial Revolution until the Clean Air Act in the 1960s, pollution in the industrial north of England darkened the trees. Against this sooty background it was the white moths which stood out, and as a result the dark variant became much more common during this period, as more white moths were removed by predation with each succeeding generation.

Such observations fit in well with Charles Darwin's view of how evolution works. But this view was the product of particular need. When Darwin published his *On the Origin of Species* in 1859, the prevailing belief was that each species was fixed, unchanging from the way in which the Creator had made it. To establish the central proposition of evolution, which is that all living things are related by descent from a common ancestor, Darwin had somehow to show that species were not fixed. His brilliant solution was to

The peppered moth of northern England is the classic example of how adaptations can become rapidly outmoded by the environment, with consequent changes in the population. During normal times pale forms of the moth are hard for predatory birds to see against tree bark, and dark forms are rare. When pollution darkens the trees, the reverse applies.

Illustration by Diana Salles.

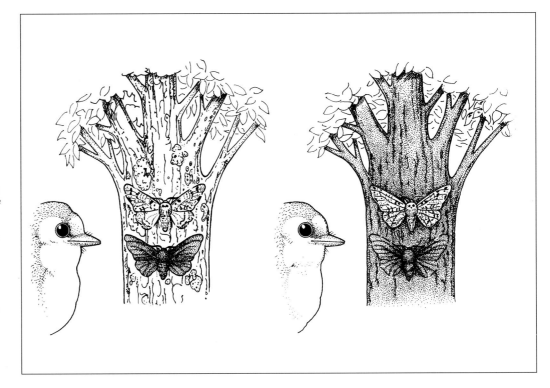

Two views of how evolution occurs. On the left we see the traditional view, in which change takes place more or less as a function of time, as species adapt to changing environments. The model on the right takes account of the fact that the fossil record reveals change more commonly to be episodic: once established, most species tend to remain relatively unchanged and to be replaced ultimately by others, to which they themselves may have given rise.

Illustration by Diana Salles.

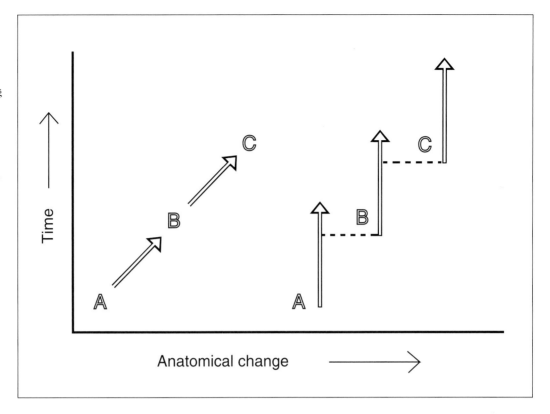

make them changeable in time, one species gradually changing into another under the guidance of natural selection. How species actually originate was, despite his title, not something that Darwin deeply explored. Traditionally, then, evolution and natural selection have been seen as simple results of differential reproduction, and change becomes inevitable given enough time. In fact, most natural selection will be stabilizing, simply pruning off unfavorable variants that occur (unusually large or small individuals, for example, tend to have lower life expectancies than those of average size); but as long as some individuals function better than others, as invariably happens, slow change is seen as inevitable.

Yet evolution does not only involve anatomical change in successive generations. A hallmark of the history of life has been the turnover of species, with the emergence of new species being balanced over the long run by the extinction of old ones. Nonetheless, a variant of the slow-change idea has been used to explain the origin of new species as well, and was popularized in the "New Synthesis" of evolutionary theory that emerged in the 1940s under the influence of such seminal thinkers as the geneticist Theodosius Dobzhansky (1900–1975), the systematist Ernst Mayr, now of Harvard University, and the paleontologist George Gaylord Simpson (1902–1984). The idea, like the one from which it was derived, is appealingly simple. From time to time, populations of any widespread species are disrupted by natural phenomena—a mountain chain builds, a river changes course. If such a barrier cuts a population in two, the two now-separate populations will each continue to change, until in the end so many differences will have accumulated between them that they can no longer interbreed. A new species is born.

The trouble is that, as we've come to realize in recent years, this simple process of long-term change doesn't fully explain the patterns we see in nature. First, the fossil record generally fails to provide the evidence of gradual change in lineages of organisms that this mechanism would lead us to expect. In fact, in general it suggests the opposite:

species, whether of mammals, or invertebrates, or anything else, tend to appear in the fossil record abruptly, to linger around for longer or shorter periods of time without changing much, and then to disappear. What's more, everything we know about patterns of ecological change in the past tells us that environments don't change gradually, at the snail's pace with which adaptation by natural selection could keep pace.

Instead, as during the Ice Ages that produced our own species, environments tend to alter erratically and rapidly, and thus the origin of new species as well as the extinction of old ones may be due to external causes that have nothing whatever to do with their adaptations. For instance, the fossil record is littered with examples of mass extinctions that took place as a result of climatic deteriorations that made the adaptations of previously successful species irrelevant to the new conditions. Yes, on balance the fossil record does reveal change over time; but the pace of the change we see probably has much more to do with changes in the rates of turnover of species than with the intensity of natural selection within them. So species, with their origins, life spans and extinctions, tend to act as distinct entities, rather than blurring one into the other over time. Like human parents they give birth to offspring but do not change into them.

What's in a species? Paleontologists have faced one major difficulty ever since their science was founded in the eighteenth century. This is that the origin of new species doesn't necessarily involve changes in anatomy of the sort that we might pick up in the skeleton—which is all that fossilizes. Speciation, the process by which new species arise, involves the establishment of reproductive isolation between populations that originally belonged to the same species. Speciation occurs when a genetic change of some kind makes the members of a population unable to interbreed successfully with others outside that population. It is a genetic event—still poorly understood today—that does not necessarily have anything to do with anatomical change. In mammals it seems that to achieve genetic incompatibility a single population must be physically divided into two parts by a geographical or ecological barrier. Anatomical change, on the other hand, originates within freely interbreeding populations as natural selection makes different attributes advantageous in different parts of their ranges. The upshot is that speciation sometimes occurs in the absence of appreciable anatomical change, even over long periods of time, while in other cases populations accumulate lots of anatomical variation without speciating.

Fossil mammals, of course, have long ceased to breed. All we have to go by in deciding what species they belong to is their age, their locality, and their bony anatomy. Since species change and anatomical change are not the same thing, disagreements frequently arise over whether a particular aggregation of fossils contains the remains of one species or more—as we will see later in this volume.

Such considerations have led in recent years to a new view of the evolutionary process that is gradually finding its way into paleoanthropology. In 1972 Niles Eldredge, of the American Museum of Natural History, and Harvard's Stephen Jay Gould published a landmark paper in which they introduced a theory of the evolutionary process that they called "punctuated equilibria." They pointed out that much of the fossil record is marked by periods of nonchange as species appear, persist unchanged (sometimes for 5 million years or more), and are finally replaced by other species. Exactly how new species (which are groups of organisms that can reproduce only among themselves) originate is poorly understood, but in geological terms the emergence of a new species is a short-term event, whether it takes five hundred years or fifty thousand. Says Eldredge: "It seems that evolution is rather like the life of a soldier: long stretches of boredom punctuated by moments of stark terror!"

Finally, it seems to me, in recent decades scientists have had a marked tendency to overestimate the amount of variation that occurs within single fossil human species. This has led them to underestimate the number of species in the human fossil record, by dismissing many real species as mere variants of others. By con-

The life history of a fossil. After death, most vertebrates will be devoured by predators or scavengers (top left). What is left over will either weather away or become buried in accumulating sediments (top right). In appropriate conditions, such buried remains may be fossilized, their organic constituents replaced by minerals (bottom left). If erosion wears away the sediments above it the fossil will be exposed at the earth's surface (bottom right), where a paleontologist must find it before it is destroyed by the elements.

Illustration by Diana Salles.

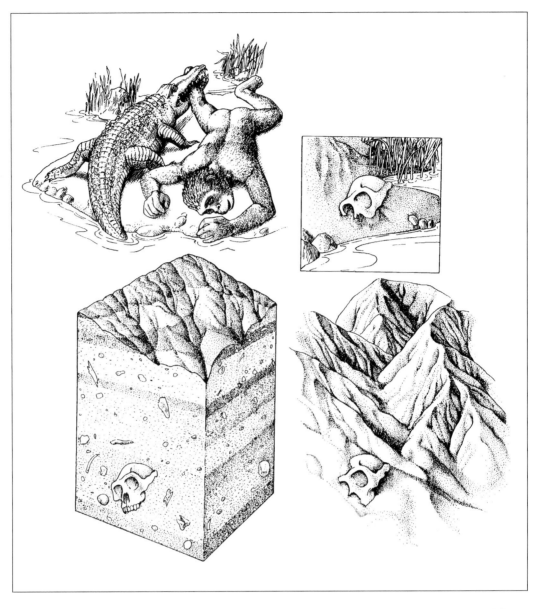

ventional reckoning, only about five species stand between us and our earliest bipedal ancestors, whereas the anatomical gulf that separates us from them is enormous. Nonetheless, some paleoanthropologists want to reduce this paltry number even further; in 1992 the University of Michigan's Milford Wolpoff and the Australian National University's Alan Thorne revived the suggestion that the archaic form *Homo erectus*, should be combined into our species *Homo sapiens*. Such measures, however, result in the spurious identification of long-persisting lineages, and a low estimate of the numbers of species that both appear and go extinct in the fossil record. This in turn disguises the real significance of the long-term trends (the increase in human brain size is the most obvious one) that are so evident in the human fossil record and that are used to validate the idea that evolutionary change is gradual. For rather than resulting from perfecting adaptation within slowly changing lineages, these trends probably result from ongoing processes of competition between related species.

Irrespective of their theoretical viewpoints, though, there is one thing on which all evolutionary biologists agree: evolution is not a goal-oriented process, however tempting it is for us to see ourselves as proceeding toward perfection, or even as having achieved it. Human evolution, like that of every other species, has resulted from a unique and opportunistic series of interactions by our precursors with an environment which included many now-vanished species and which fluctuated in a pattern which will never exactly repeat itself. If we were to replace ourselves today with a remote ancestral species and start the whole process all over again, there is no guarantee that several million years hence another *Homo sapiens* would stride the earth. In fact, it would be exceedingly improbable.

What are fossils?

Our species, then, is the result of a unique series of evolutionary events. And even though we can recognize our relatives of varying degrees by comparing our structure with those of the species around us, it is only the fossil record that can reveal the precise details of our biological history. So just what are fossils? Technically, a fossil may be any evidence of past life (such as footprints or the impressions of tree roots or wasps' nests); but in practice the human fossil record consists almost entirely of bones and teeth. These are the most durable tissues of the body, those most likely to be preserved.

It's not easy to become a fossil. The process starts at the death of an animal, when its carcass is immediately at the mercy of predators and scavengers. These creatures will usually dismember it, consuming or carrying off various body parts, and this is one of the

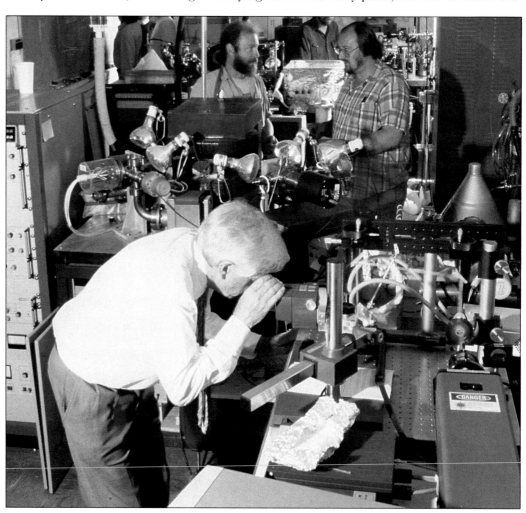

View of the Geochronology Laboratory at the Institute of Human Origins, Berkeley, California.

Photo courtesy of Institute of Human Origins.

major reasons why complete fossil skeletons are rarely found. In fact, fossils are seldom discovered where their owners fell; more commonly, they turn up in places where water or other animals deposited them.

Assuming that they have not been entirely eaten, the bones must then escape the destruction of weathering until they are covered by accumulating geological sediments. These sediments, which will later turn into rock of varying hardness, must be of a kind that is not destructive to bone; they should not be too acidic, for example. The best place to look for the fossils of terrestrial vertebrates, from dinosaurs to humans, turns out to be in rocks formed from the relatively rapid buildup of sediments at lake margins or on the floodplains of rivers. Once bones have become incorporated into the accumulating geological strata, the rocks in which they reside cannot undergo too much modification, such as fracturing or melting due to deformation or pressure; such processes obviously destroy fossils. During their sojourn in the rocks, the original materials of which the bones and teeth were composed are replaced to a greater or lesser extent, generally by minerals that infiltrate them in water that passes through the sediments.

The sedimentary rocks of the world are packed with fossils; but to be of any use to paleontologists, of course, the fossils have to be found. For this to happen, erosion has to cut down through the rock in which they are encased and expose them at the earth's surface—at which point the fossils themselves will start eroding as they are exposed to wind and weather. So finally someone who knows how to recognize a fossil needs to come by at the critical moment when the fossil is visible on the surface, but has not yet been destroyed by the elements. Small wonder, then, that only the tiniest fraction of all the vertebrates that have ever lived are known from fossils!

Dating rocks and fossils

Fossils provide an important key to understanding the history of the Earth. This importance stems from the fact that sedimentary rocks (those composed of compacted wind-borne or waterborne particles weathered from pre-existing rocks and deposited anew) form a large part of the geological record, but generally cannot be traced physically over long distances. The fossils they contain, however, are characteristic of particular periods. Such fossils, often called index fossils, can be used to assign rocks to their appropriate place in the long sequence of geological events. This is far from an exact process, but it did allow the development of a worldwide system of relative dating (older than this, younger than that) long before it became possible to assign year dates to rocks using completely different methods. Fossils are also useful in the reconstruction of ancient environments; not only can we tell from the rocks themselves what the local environment in which they were deposited was like, but because different animals are characteristic of different environments, the group to which a fossil animal belongs can help us to infer the environment in which it lived.

In the years since World War II a number of new techniques have been developed that allow us to assign year dates to certain kinds of rock. In paleoanthropology the most familiar technique of this kind is the potassium/argon (K/Ar) method, first applied to a human fossil site (the famous Olduvai Gorge, in Tanzania) around 1960. This method depends on the breakdown of a radioactive (unstable) form of the element potassium (^{40}K) to an inert form of the rare gas argon (^{40}Ar). Such breakdown occurs at a constant rate. Thus, if we measure the amount of ^{40}Ar in a rock sample, compare it to the amount of radioactive potassium that occurs in newly formed rock, and apply the rate of breakdown, we can arrive at a year estimate of the time that has elapsed since the rock formed.

Though there are numerous practical pitfalls in a method as complex as this, volcanic rocks have been a particular favorite for K/Ar dating for two reasons. First, they crystallize at high temperatures at which all argon originally present will have been driven off; and second, they are excellent "stratigraphic indicators." Since volcanic rocks—lava flows, ashfalls, and so forth—descend on the landscape at a particular instant in time, a layer of volcanic rock will usually be just slightly younger than the rocks that lie beneath it and slightly older than those that accumulate on top of it. So even if the actual fossils themselves cannot be dated in this way, a datable lava flow that lies just above or below a

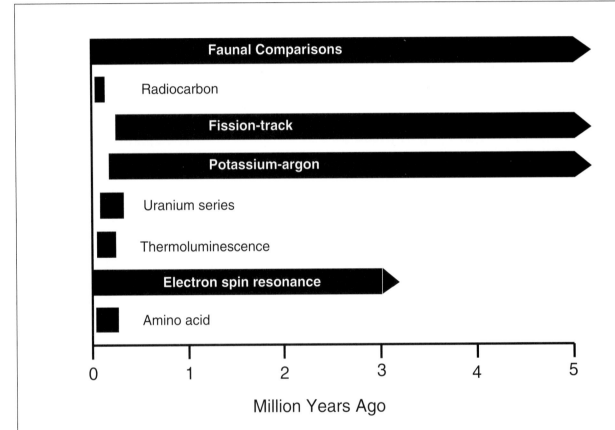

Million Years Ago

Methods of dating fossils. Alongside faunal dating and the longer-established methods of year-dating such as carbon-14 and potassium/argon, described in the text, several new methods have recently come on the scene. Among them are fission-track dating, in which the damage done to mineral crystals by decaying radioactive atoms is quantified; uranium-series dating where the decay products of various isotopes of uranium (particularly in carbonate rocks such as calcites) are measured; thermoluminescence, in which the light emitted by released electrons under heating reflects the time since various minerals (e.g., flint) were last heated; the somewhat similar method known as electron spin resonance, which also depends on measuring free electrons trapped in solids—and which has the attraction that it can be performed in certain cases on dental enamel; and amino acid dating, which uses the tendency over time of amino acids in organic objects to change the direction in which they are coiled. This last form of dating is a rather approximate procedure, most useful as a confirmation of other methods, although it has been claimed that materials such as ostrich eggshells yield quite accurate results. Electron spin resonance and thermoluminescence have the drawback that they require information about past climates that may not always be available. The diagram above shows over what periods of the past these various methods of dating are applicable.

layer containing fossils will give a good estimate of the age of those fossils. Because the breakdown time of ^{40}K is very long (over a billion years), the technique is best used on older rocks above about a quarter-million years in age, though it has been used on younger rocks.

Various other techniques exist that work on similar principles, and some are becoming available that can be used on certain kinds of sediment. Yet others can sometimes be used on fossils themselves, though generally not on very old ones. The best-known such method is the radiocarbon (^{14}C) technique. Here's how it works: All living beings contain carbon. Most of this carbon is stable, but one type (^{14}C) is radioactive. As long as a creature remains alive the ratio of stable to radioactive carbon that it contains stays the same. But when it dies the radioactive part starts to break down and diminishes in quantity. Since the rate of breakdown is constant, the ratio of radioactive to stable carbon in organic remains will tell you how long it has been since the organism died. The breakdown time of radiocarbon is very short (under 6,000 years), however, so that after about 50,000 years there will be too little radiocarbon left to measure accurately. This places a maximum of about 50,000 years on reliable radiocarbon dating.

Recently, various methods have come into use that to some extent fill the gap between the useful time ranges of K/Ar and ^{14}C dating, and which can be used on different materials. These include thermoluminescence dating, which measures how many free electrons are trapped in an object. This method has proved particularly useful in dating archaeological sites in which artifacts were burned in cooking fires. Despite such advances in "chronometric" dating, there is still a pretty good range of situations in which methods of this kind cannot be used. In these cases we are forced back to the more traditional approach of dating fossils by comparisons with faunas elsewhere.

Dating is, of course, only one aspect of the study of fossils, simply placing them within the sequence of evolutionary events. Comparisons with other forms, both living and fossil, allow us to determine to which other species they are most closely related. Functional studies of their anatomy help us to understand how they behaved in life, and the picture may be more fully fleshed out by reconstructing the environments in which they lived. In the case of humans the archaeological record, which documents the development of technological achievement and cultural complexity, adds an important dimension to the story. By incorporating all of these approaches, modern paleoanthropology strives to develop as complete a picture as possible of the lives, behaviors, environments and evolutionary relationships of our forebears.

The background to human evolution

Although we resemble our relatives the great apes in our basic body structure, we are more specialized in a number of features. We have, for example, a large and internally reorganized brain; we walk upright on two legs, which has involved considerable modifications throughout our body skeleton; we have reduced faces and teeth (particularly the canines); and we have an unparalleled dexterity which we employ to make and use sophisticated tools. It is generally agreed that any hominoid possessing one or more of these features may be placed in the human family. Hence human evolution is largely the story of the acquisition among our various precursors of these particular bodily structures and abilities.

Naturally the term *human* has been used since long before anyone had any idea that our species has an evolutionary record that ties us to other humanlike species. So when we use this adjective to describe ourselves we usually have in mind those of our qualities that we prize most, such as high intelligence, aesthetic sensibilities, and language. But what of those precursors in our lineage who shared some, but not all, of these and other features with us? Or in whom they were less highly developed than they are in ourselves? At what point in the history of human divergence can we say that those precursors became human in a true functional sense? There is no objective answer to this question. Certainly, it seems from the archaeological record that *fully* human sensibilities only emerged after about 50,000 years ago, long after the appearance of anatomically modern people. But it is nonetheless quite possible that were we to meet various of the more archaic human types in the flesh we might see enough of ourselves in them to recognize them intuitively as human.

This leaves a genuine gap in terminology, for it is technically no longer possible to use the familiar word hominid to denote the group that contains just us and our close fossil relatives. For since most scientists now classify the great apes along with us in the family Hominidae, we cannot use the derivative term hominid to identify any narrower group within it. In this book I compromise. I use the term human to denote, rather than to describe, the group of hominoids whose evolution is described in the remaining chapters.

Chapter Six

The Earliest Human Relatives

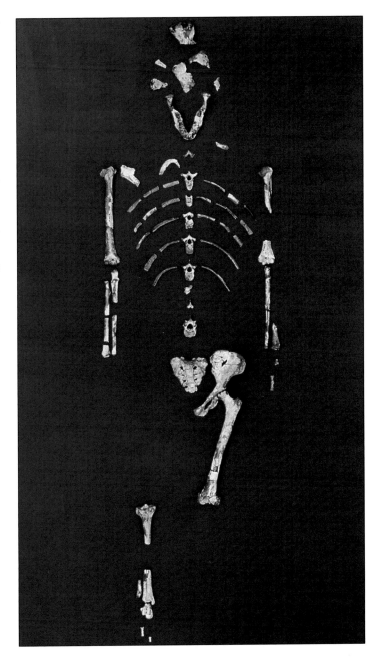

As we've already seen, evolutionary history is often shaped by ecological events unrelated to how well a species is adapted to the environment in which it has been living. For our species, the most momentous event of this kind was the increase in seasonal aridity that struck Africa about 7 million years ago, causing a major expansion of open habitats and a corresponding reduction in forests. For it was out of this climatic and ecological stress that a new kind of hominoid primate emerged some time before about 5 million years ago, as the Miocene ended and the Pliocene epoch (5–1.6 million years ago) began. This hominoid, adapted to life in more-open country, was our earliest ancestor.

Photograph of "Lucy" (NME AL 288-1), the famous skeleton of *Australopithecus afarensis* discovered at Hadar, Ethiopia, in 1974. Approximately 3.2 million years old, "Lucy" is about the youngest, as well as the smallest, of the specimens attributed to *A. afarensis*.

Photo courtesy of the Institute of Human Origins.

The first upright walkers

A very poor fossil record from some 8-4 million years ago makes it difficult to know exactly when and how this new primate emerged, although there's no doubt about the place: Africa. The oldest fossils that document the fact of human emergence are a few bits and pieces about 5 million years old that come from sites in Ethiopia and Kenya, in sediments associated with the evolving Rift Valley. By themselves they don't tell us much, but their age gives them some importance because they appear to extend back the range in time of the kind of early human that is best documented from the localities of Hadar, in northern Ethiopia, and Laetoli, in northern Tanzania.

The first of these sites is most famous as the place where the skeleton "Lucy" was found late in 1974 by Donald Johanson, founder of the Institute of Human Origins in Berkeley, California. Lucy consists of about 40 percent of the bones of a very small-bodied individual, who stood in life about three feet, six inches tall and is presumed to have been female. Given the astonishing rarity of early human skeletons, Lucy's discovery was one of the most remarkable paleoanthropological finds ever. It was a fitting climax to weeks of mind-numbing clambering over rough, shadeless desert in the unrelenting sun, and by Johanson's own admission, it revolutionized his existence.

Lucy is, however, only one of a large number of early human remains found at Hadar since 1973 by Johanson and his colleagues, and at approximately 3.2 million years old she is about the youngest, as well as the smallest. Some of the Hadar specimens date up to a couple of hundred thousand years earlier, and the whole collection includes such other notable fossils as the "first family," a trove of bones from a single locale that contains parts of at least thirteen individuals who may have perished together in one single catastrophic event, perhaps a flash flood.

Approximately 3.6 million years old, the main sites at Laetoli are a little older than the oldest ones at Hadar and have yielded a much smaller collection of early human bones, mainly a few partial jaws. The jewel of Laetoli is, however, a unique trail of fossil human footprints several yards long, preserved in a bed of volcanic ash along with trails left by other creatures.

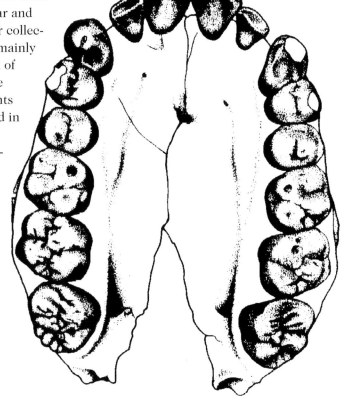

A palate of *Australopithecus afarensis* (NME-AL 200-1a) from Hadar, Ethiopia, dated at about 3.4 million years old.

Illustration by Donald McGranaghan.

Part of one of the foot-print trails at Laetoli, Tanzania, dated to about 3.5 million years ago. Ancient as these foot-prints are, they show a perfect two-footed stride. At least two individuals, one large, one much smaller, strode across wet volcanic ash at Laetoli to leave these footprints behind; they must have walked togeth-er since their stride lengths are matched to each other, and, if so, they must have been close enough to have been touching.

Photo courtesy of Peter Jones.

The footprints at Laetoli

Three and a half million years ago, the nearby volcano Sadiman puffed out a cloud of fine ash over the landscape around Laetoli. Falling rain then turned the ash to something like wet cement. Over this soft surface walked at least two early humans, leaving behind them footprints just as one does on a wet beach. Later the volcano spoke again, covering the prints with more ash and preserving them as part of an accumulating pile of sediments. These sediments much later eroded and eventually the footprint layer was exposed again at the surface of the undulating Laetoli landscape, where it was discovered in 1976 by members of a team led by Mary Leakey. What an incredible find!

Perhaps the most interesting thing about any extinct animal is how it behaved. Yet usually behavior has to be inferred indirectly from the evidence of bones and teeth, and there is almost always argument over inferences of this kind. But at Laetoli, through these footprints, behavior itself is fossilized. And the most astounding thing is that these ancient individuals were walking upright *just like us*. The Laetoli foot-prints are astonishingly similar to those that barefoot modern people make on muddy paths. They are so similar, in fact, that some scientists have had difficulty in believing that they were made by the same early humans whose bones were found at

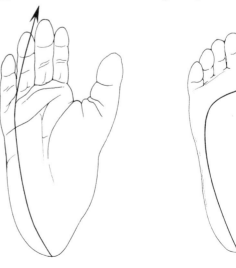

Contrast between the ways in which weight is transmitted along ape and human feet. In an ape moving bipedally, the weight is carried along the side of the foot, and on toe-off is passed through the middle of the row of toes. In a human, body weight is transmitted along the outside of the foot, then internally across the ball of the foot, and finally push-off is made by the big toe—a much more efficient arrangement for striding.

Illustration by Diana Salles.

Reconstructions of the early humans who made the Laetoli footprints, as shown at the American Museum of Natural History. These
figures are based on fossils attributed to *Australopithecus afarensis*,
but although their body proportions are thus presumably accurate,
many details of these reconstructions are entirely conjectural.
Among the attributes that can only be guessed at are hair density
and distribution, skin color, form of the nose and lips, and many
other features.

*Sculptures by John Holmes, photo by Dennis Finnin
and Craig Chesek*

Laetoli and Hadar. Yet there are
no other known candidates
from the time the footprints
were made 3.5 million years
ago. And there's little room for
quarreling over how those who
made these footprints walked.
There is a fundamental difference in the way weight is transmitted along the feet of modern
bipedal humans and of
quadrupedal apes (even when
they walk on two legs), and
there's no confusing the tracks
they leave behind, at least
when there are many of them
to go by.

Southern apes of the Afar

The Hadar and Laetoli human remains are classified by most scientists in the species
Australopithecus afarensis (the name literally means "southern ape of the Afar," the
region of Ethiopia in which Hadar lies), though there's debate about whether all of the
specimens represent individuals of the same species. If they all do belong to the same
species, there is an enormous difference in size between the largest (presumably male)
and the smallest (presumably female) individuals. Size differences of varying degrees
between males and females are, however, pretty much standard for large living hominoids. Females of *A. afarensis* stood about three feet, six inches tall and weighed on average perhaps sixty pounds; males reached maybe four feet, six inches with some weighing
well over one hundred pounds. Even before the discovery of the Laetoli footprints, scientists knew that these early humans walked upright on two feet. They knew this because
among the first human fossils to be discovered at Hadar was a knee joint that showed all
of the characteristics of upright walking.

Nonetheless, the body skeletons of these early humans were very different from
ours, most notably in their limb proportions. Lucy, for example, shows clearly that while
the arms of *A. afarensis* were proportioned pretty much like ours in relation to the trunk,
the legs were rather short. Other specimens show that the hands and feet were proportionately longer than ours, with some curvature in the bones, indicating that these early
humans had retained some climbing abilities. This makes sense, for the earliest humans
would not be expected to have lost all traces of their arboreal ancestry. Moreover, the
open savanna is not a very safe place for small-bodied and relatively slow-moving animals
such as our ancestors; it is logical that they sought the shelter of the trees for sleeping at

Aerial view of the fossil-yielding sediments at Hadar, Ethiopia. Today a thin strip of gallery forest along the Awash River yields to desert-like badlands; 3 million years ago a stream-fed lake existed here, its forested margins giving way to woodland and open savanna.

Photo by Donald Johanson; courtesy of the Institute of Human Origins.

night or for resting during the day. Though we should not underestimate their toughness–for their robust skeletons show that these early relatives were very muscular, and the larger ones among them were as big as chimpanzees, which are formidable adversaries—there's no doubt that newly bipedal creatures were vulnerable on the African savanna. But it's also possible that the retention of some arboreal features by *A. afarensis* was related to food gathering. It has even been suggested that early humans gained much of their sustenance from climbing trees to steal the carcasses that leopards probably left unattended there for hours at a time.

Determining how smart these ancestors were is a matter of not-very-well-educated guesswork. Brain size is only a very rough indicator, because how the brain is organized internally is at least as important as how big it is. Nonetheless, it is a measure of sorts. Modern human skulls have a large rounded braincase with a small face tucked under the front end; the skulls of *A. afarensis*, in contrast, showed the opposite, typical hominoid, condition with a large, projecting face and a small braincase, at least compared to our own. This contained a brain of about 400 milliliters in volume, hardly larger than that of a modern chimpanzee and under a third of the modern human average of about 1,350 milliliters. It was, though, proportionately much larger than that of, say, *Proconsul*.

There's also no direct evidence that early humans used tools until well after the time of *A. afarensis*. Of course, only the introduction of durable stone tools would have left such proof, and it's quite possible that *A. afarensis* used softer materials like wood for tasks such as digging up roots and tubers; after all, chimpanzees are well known to go "fishing" for termites with stripped twigs and otherwise to modify natural objects for useful ends. Without an archaeological record to go on, however, we lack most of the elements that we need to make inferences about the behavior of a hominoid that has no near equivalent among living primates. Most of what we can surmise about behavior comes indirectly, from what the fossil sites in which *A. afarensis* is found suggest about its habitat.

Ancestral environments

Unfortunately the Laetoli sites don't provide many clues to lifeways, since it is unlikely that their immediate surroundings were very congenial to Pliocene humans. This was an austere, windblown area that was at least seasonally extremely arid, and was covered with stunted grass, scattered acacia trees, and other thornbush. The immediate area of the human trackways was a harsh, dusty plain devoid of vegetation. "This can't have been the kind of environment in which these early bipeds normally lived," says Peter Jones, one of the archaeologists who excavated the prints. Jones thinks that the individuals who made the tracks were just passing through. "Maybe," he surmises, "they were heading for the more hospitable Olduvai basin, just over the horizon, where a lake lay in the middle of a wooded depression."

Fortunately, Hadar gives a much better idea of the environments in which *A. afarensis* may actually have lived. A lake lay there, supplied by water draining off the edge of the Ethiopian Rift. The lake fluctuated in size, but its environs ranged from forest to woodland and open savanna. Most of the fossils recovered in this area are associated with lake-edge deposits, which would have been tree-fringed. This mosaic of closed and more open habitat types fits well with the spectrum of adaptations we see in the skeleton of *A. afarensis*. These early humans probably sheltered and fed in the forested areas, emerging into the more open grasslands to scavenge and maybe also to grub for roots and tubers. It's unlikely that they actively hunted anything much larger than the very smallest game, though chimpanzees, for instance, do cooperatively hunt small mammals such as monkeys and baby antelopes—even other chimpanzees.

In any event, the environment at Hadar was close to what we would expect for a hominoid that had taken the first major step toward becoming an open-country-living creature, but which did not show all of the bipedal characteristics that humans later acquired. Why this step toward an open-country existence was taken has been debated, but presumably it represented a way of coping with an environment in which forests were becoming scarcer. What the social consequences of this change may have been we'll never know for sure, though it may be significant that all living higher primates are gregarious, and that groups of them are tied together most importantly by the mother-offspring (and particularly mother-daughter) bond. This bond might well have provided the principal stabilizing element in the groups of *A. afarensis*, and most especially the basis for larger kinship networks. Within the group, communication remained relatively unsophisticated; certainly anything like language was far in the future. Groups would probably have remained small, moving around constantly within a large territory, and perhaps coming together during seasons when localized food resources were particularly abundant.

Side view of the skull of *Australopithecus afarensis* compared to that of a modern human. The fossil skull is reconstructed from fragments belonging to various different individuals, and the comparison shows the enormous difference in the proportions of the face relative to the braincase between the earliest and the most recent members of our lineage.

Illustration by Diana Salles.

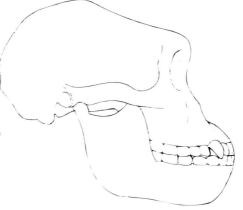
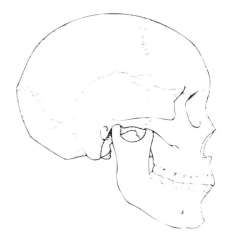

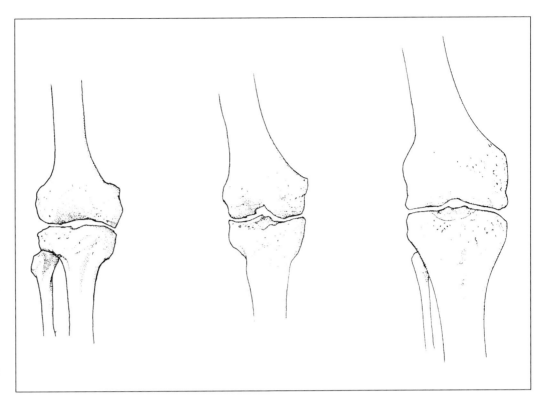

Contrast between the knee joints of a modern human (left), a chimpanzee (right), and *Australopithecus afarensis* from Hadar. The Hadar and modern knee joints resemble each other in having a "carrying angle," typical of human bipeds. In the chimpanzee the thigh and shin bones form a straight line, which is an ideal arrangement for a form that bears its weight on all four limbs. However, when walking on two legs this means that at each pace the body's center of gravity must be swung inefficiently in a circle around the supporting leg. In contrast, in the human and Hadar forms the thigh and shin bones form an angle, so that the thighs come together at the knees, and the feet trace a single straight line.

Illustration by Diana Salles.

The anatomist Owen Lovejoy has argued eloquently for the formation of permanent exclusive bonds among pairs of *A. afarensis* males and females within the group. This is likely, he feels, because a female's reproductive success is determined by how many offspring she can support at one time, and because this number is increased by help from a male, especially in an environment where food resources are very scattered. Unencumbered by infants, the males could move around more freely than females, and pair-bonding would ensure that the offspring that they were helping to support were their own. It would also help to lower tensions within the group, which needed to stay together because of the safety of numbers it provided when out on the open savanna—though this is not the only way of achieving such a result: bonobo chimpanzees seem to relieve group tension via virtually indiscriminate sexual activity among all adult group members. Lovejoy has suggested further that such a development may also explain the origin of bipedalism, since walking on two legs would free the arms of males to carry food home to the group. "If your mate is walking upright," Lovejoy says, "he's better equipped to carry food, and more likely to bring some to you."

Why upright posture?

As the first major unique characteristic to have been acquired in the human line, and the one that laid the groundwork for all subsequent developments in human evolution, the reasons for the adoption of bipedalism have been debated loud and long. Lovejoy's association of bipedalism with pair-bonding is far from the only possible explanation for it; indeed, it has been widely attacked on a number of grounds. For example, *A. afarensis* shows a great disparity in size between males and females. In primates generally, this sex difference is associated with competition among males for females, and not with pair-bonding at all. One competing (though not uncontested) explanation for the adoption of upright walking is quite simply that for a hominoid (as opposed to a more thoroughgoing four-legged form such as a monkey), bipedalism is at least as efficient in open environ-

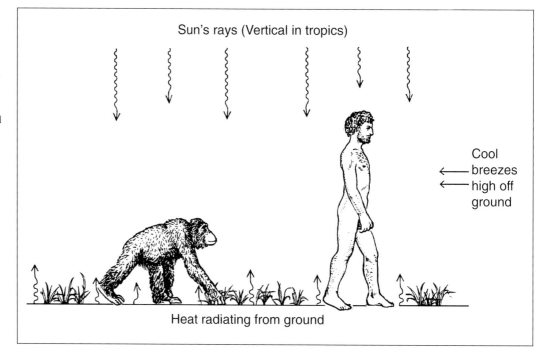

Some of the conse-
quences of a bipedal
versus a quadrupedal
posture in the unshaded
tropical savanna. Com-
pared to the quadrupedal
ape, the bipedal human
reduces the area of the
body receiving both the
direct rays of the sun and
the heat radiating from
the earth. The bulk of
the body is also raised
high off the ground, thus
benefiting from the cool-
ing effect of the wind.

*Illustration by
Diana Salles.*

Sun's rays (Vertical in tropics)

Cool
breezes
high off
ground

Heat radiating from ground

ments as is going on four legs. Indeed, it is considerably more so once the basic human
adaptations to two-legged locomotion have been acquired; with diminishing forests came
the need to cross over open ground between patches of trees, and efficient bipedalism has
been seen as an obvious means of accomplishing this goal.

More intriguing, though, is the theory advanced by the English physiologist and
paleoanthropologist Pete Wheeler, who suggests that accommodating to the open environments which replaced their forest habitats exposed the ancestral humans to a problem
that their precursors had never had to face. This was the heat load imposed on them by
the direct tropical sun. The body both produces heat and receives it from the environment, and in open tropical conditions there is a real danger of overheating and heat-
stroke. A way is needed of shedding body heat while at the same time reducing its absorption. A quadruped absorbs heat over much of its body, but by standing and walking
upright the area of the body that receives the vertical rays of the overhead sun is reduced
to the crown of the head and the shoulders. The rest of the body, more sheltered from
those rays, is available to lose heat by radiation, convection, and other means.

This explanation for the adoption of upright posture has the added attraction of
invoking another of our unique qualities: our hairlessness which, in association with an
extensive distribution of sweat glands, gives us the most effective body cooling system of
any mammal. We are descended from a hairy hominoid; and, in fact, in the course of our
evolution we have not reduced the number of hairs we have, just the size of nearly all of
them. Dense hair impedes heat loss, directly by radiation and, more importantly, by the
evaporative cooling of sweat. The adoption of bipedalism enhances heat loss in an animal
that actively moves around on the tropical savanna not simply by reducing the area of
the body that absorbs heat, but by raising the body high above ground level, where higher
wind speeds enhance cooling by the evaporation of sweat. "Stand tall," wrote Wheeler,
"and stay cool." For an already potentially bipedal hominoid, then, bipedalism and hair
loss appear to be a natural combination of adaptations to life out on the tropical savanna.
This remains true even if initially the main purpose of being out in the open was to move
between areas of shrinking forest.

Was Lucy really our ancestor?

Whether or not it is related to later events such as brain expansion, bipedalism is certainly the original human adaptation and it is the major reason why _A. afarensis_ is accepted within the human clan. But it's hardly surprising that, once the initial excitement of discovering the earliest human died down, it became popular to describe this species as an ape that walked upright. For its skull has, like the living great apes, a large chewing apparatus grafted onto the front of a relatively small braincase. Its face projects forward because of the great size of its teeth, both behind and in front of the canine. Indeed, a back-of-an-envelope calculation suggests that a modern person who had molars as large as those of this four-footer would stand more than twice as tall. But the canine tooth, while larger than in later humans, is considerably smaller in _A. afarensis_ than in any great ape. We can thus consider the dentition of this species to be roughly intermediate between those of modern humans and modern apes.

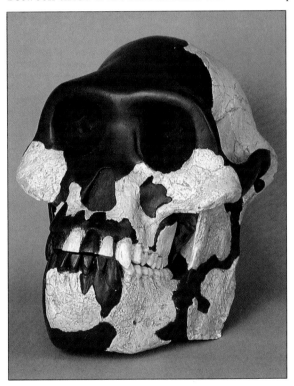

What does all this imply about the evolutionary position of _A. afarensis_? Most scientists believe that this species is the ancestor of all later humans, or at least a very close relative of that ancestor. However, some see more than one species in the aggregation of fossils assigned to _A. afarensis_. And while this is very much a minority view at the moment, it has great potential for revival. Most scientific opinion, though, is divided between those who see this _A. afarensis_ as giving rise to a descendant species that was the immediate ancestor of the various later branches of the human family tree, and those who see these branches as all stemming (in various combinations) directly from _A. afarensis_. We'll look at this complex picture later.

The skull of _Australopithecus afarensis_, as reconstructed from fragments of different individuals collected from sites at Hadar, Ethiopia.

Photo by Willard Whitson,
Illustration by Don McGranaghan.

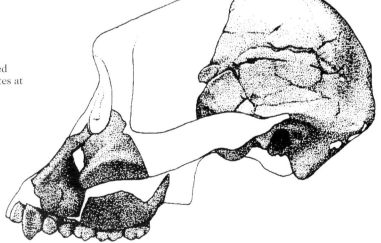

The southern ape of Africa

The first truly primitive early human precursor to be discovered was *Australopithecus africanus* (literally, "southern ape of Africa"), described in 1925 by Raymond Dart (1893-1988) at a time when the human fossil record was very limited and contained nothing that was in the least comparable. As he was dressing to be best man at a friend's wedding, Dart, an anatomist at a South African university, was given a box of fossils from a lime mine at Taung, a remote spot on the fringes of the Kalahari Desert. Fossil baboons were known from the site, but among the specimens delivered to Dart was a natural cast of the inside of a braincase, formed from sediments that had filled the inside of a skull. This brain was far too big to come from a baboon. Further inspection of the shipment of fossils produced a block of compacted sediment into which the front of the brain cast fitted, and after much painstaking labor on Dart's part, this block yielded the face of a young individual.

It was this specimen, the face and brain cast of a child at about the stage of development of a modern six-year-old (maybe younger), that Dart named *A. africanus*. Despite the name he gave it, Dart described this specimen as intermediate between apes and humans. Because young humans and young apes resemble each other much more closely than adults do, and because this specimen was so different from anything previously seen, it is not surprising that Dart's interpretation ran into a lot of resistance from the scientific community—which at the time already had its hands full trying to figure out where the famous Piltdown "fossil" fitted into the picture of human evolution. Piltdown, of course, was later shown to have been a hoax, consisting of broken bits of a modern human braincase and the jaw of an orangutan, cleverly modified to disguise the mismatch. But between 1912 when it was first "discovered" and 1953, when the fraud was exposed, Piltdown did much to derail researches into human evolution, especially in England, where many of the most influential paleoanthropologists of the period worked.

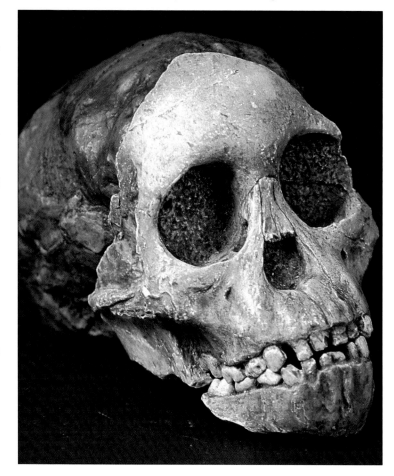

The skull of the "Taung child," the first specimen of *Australopithecus* to be described (in 1925). The face and natural brain cast of this young individual, equivalent to a modern five- or six-year-old, found by Raymond Dart in a box of miscellaneous fossils discovered by lime miners near Taung, in central southern Africa. Dart's claim that this specimen represented an intermediate form between apes and humans took decades to become generally accepted.

Photo by Willard Whitson.

Recent view of the now-abandoned lime-mining site near Taung, southern Africa, which produced the first *Australopithecus* fossil. The breccia deposit that yielded the actual fossil was long ago destroyed by the mining operation; we see here the remnants of the limestone cliff that was mined.

Photo by Willard Whitson.

Dating the southern Apes

Dart's original reading of the Taung specimen turned out to be fully justified when later finds of adult specimens were made. These came from the sites of Sterkfontein and Makapansgat, also in north-central South Africa, in the years following the first discoveries at these localities in 1936 and 1947, respectively. Before we look at the fossils, however, let's briefly consider the nature of these three sites, and others like them, for they are unusual places in which to find early human fossils.

The surface rocks of much of central South Africa are ancient dolomitic (magnesium-rich) limestones which, like all such rocks, may be dissolved away by water action. As it dissolves the rock, the water becomes supercharged with lime. This lime, carried along in the water, is then available to be redeposited in a particularly pure form. The underground cavities which form in the limestone as water dissolves it are thus gradually refilled by redeposited lime. Such cavities can also be filled by miscellaneous rubble if, again by solution, a shaft forms that connects the cavity with the surface. Material that falls down the shaft and collects on the floor of the cavity includes sand, rock fragments, and sometimes bones as well, and as redeposition continues these bits and pieces become cemented together by lime into a rock-hard substance called breccia.

Continuing erosion of the surrounding limestones eventually exposes these breccias at the earth's surface. It is in exposed cave breccias of this sort that the South African early human fossils are recovered (see diagram on page 84), and all the fossil sites were initially found by lime miners who were after the pure redeposited cave limestones. In 1925 such lime was in great demand for a newly developed process of gold refining, and for the miners the breccias were simply inconveniences to be blasted away—incidentally revealing fossils in the process.

Stages in the formation of a typical South African *Australopithecus* cave site. First, an underground cavity forms in the soluble limestone while the water table is high. When the water table drops, water filtering through the rock redeposits lime on the walls of the cavity. If by further dissolution of the limestone a connection forms with the surface, rocks, dust, and bones can fall in, to be cemented into a hard breccia by more redeposited lime. Often, the roof of the cave will collapse atop the breccia. If the overlying limestone at last is eroded, the fossil-containing breccia will be exposed at the earth's surface.

Diagram by Diana Salles, after a concept by C.K. Brain.

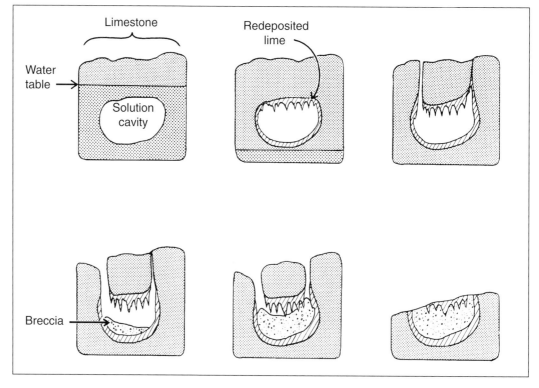

One of the problems with bones that have fallen into caves and become fossilized is that the time at which they were deposited cannot be determined from the rocks in which they are found. Sedimentary rocks normally lie in a sequence, and the fossils they contain can be dated accordingly. But there is no similar way of figuring out when a cave formed and when bones fell into it. The only way we can estimate the age of fossils in cave breccias is to compare the species contained with those known and dated from normal sedimentary rocks elsewhere.

Fairly extensive fossil faunas exist from the South African cave sites, but there are additional complications, especially in the case of Taung. The limestones at Taung were extensive, having formed not in a small cavity but as flowstones, freshwater limestones deposited over a long period as a stream (still active) passed over the edge of an escarpment. The particular breccia pocket from which the Taung early human fossil came was long ago destroyed by the mining operation, and it's far from certain that the baboons in Dart's box were from the same spot and thus of the same age as the human fossil. A recent best guess for the age of the Taung child, based on comparisons with dated faunas from East Africa, is something over 2 million years. That's plausible enough, but it is still just a guess.

Dating Sterkfontein is easier, even

Upper jaw and teeth of *Australopithecus africanus* from Sterkfontein, South Africa.

Illustration of Sts 52a by Don McGranaghan.

though the geology of the cave fill is complicated. This is because the site has been excavated mainly by paeontologists, and there are lots of early human and other fossils known from specific places within the locality. The _A. africanus_ specimens come from earlier breccias at the site, in deposits known as Member 4 and faunally dated to about 2.5 million years ago. Even earlier are the _A. africanus_ fossils that come from deposits described as Members 4 and 5 of the Makapansgat limeworks site in the northern Transvaal, which may be around 3 million years old. This gives _A. africanus_ a time range in South Africa from about 3–2 million years ago, hence in the period immediately after the heyday of _A. afarensis_ in East Africa. It has yet to be shown conclusively that fossils of this exact species have been found anywhere else.

Killer apes?

Dart's early studies of broken bones found alongside those of the early humans in the cave breccias caused him to conclude that _A. africanus_ was a hunter. Indeed, as time passed he became more passionate on the issue. _Australopithecus_, he once wrote, was a bludgeon-wielding "murderer and flesh hunter," whose violent proclivities led inevitably to the "blood-spattered, slaughter-gutted archives of human history." This was stirring stuff, and it certainly stirred the imagination of the journalist Robert Ardrey, whose speculations on the "killer ape" in _African Genesis_ and other books widely popularized the view that humanity's birth was "red in tooth and claw." But careful analysis of the bones from the cave sites shows that the remains of _Australopithecus_ that collected in the caves were much more likely the leftovers of meals of carnivores such as leopards, and of scavengers such as porcupines and hyenas. Even today, trees on the high veld of South Africa often grow next to sinkholes leading to underground cavities, and leopards repair there to consume their prey. Often, bits of the prey find their way down the holes to the spaces beneath. And as we'll see in a moment, there is direct evidence that some early humans also fell victim to leopards.

Early human lifestyles

Australopithecus africanus did not differ very conspicuously from _A. afarensis_ in its build, although a number of minor differences in the skull and teeth have been pointed out. For example, at about 440 milliliters, brain size tends to be a little larger than in _A. afarensis_, but the lower face still protrudes markedly. The body skeleton is less well known than that of _A. afarensis_, but in known elements does not differ greatly; clearly, here was another upright-walking early human with rather primitive body proportions that retained a considerable capacity for climbing as well as being an efficient biped on the ground. As with _A. afarensis_, the pelvis showed a radical reorganization from the primitive hominoid type toward the modern bipedal condition, and the spine just above it was curved forward like ours, suggesting a very similar posture. It's hard to estimate body weights in the absence of reasonably complete skeletons, but one estimate for males and females together suggests an average of a little over a hundred pounds.

Somewhat as in _A. afarensis_ there is even a substantial difference in size—particularly in the chewing teeth—between the largest and the smallest human fossils recovered from Member 4 of Sterkfontein and at Makapansgat. We should be cautious about this, however, because it does seem likely that two species of early humans are represented at these sites. If this is so, the larger ones belong to an as-yet undescribed species of rather "robust" tendencies.

The teeth of _A. africanus_ indicate that it was a herbivore. The ways in which the teeth are scratched and worn, in particular, suggest that it ate both fruits and foliage. There's not much direct evidence for eating meat, although small vertebrates probably

View of the site of Makapansgat in the northern Transvaal of South Africa, which has produced important fossils of *Australopithecus africanus*. This vast cave was excavated by lime-miners, in whose rubble dumps early human fossils were first found by James Kitching in 1947.

Photo by Ian Tattersall.

provided food at least occasionally. The Sterkfontein site indicates that the Member 4 deposits accumulated while somewhat wetter conditions prevailed than during the period in which the succeeding Member 5 fossils were deposited, and that the surrounding environment was rather bushy. Forest presumably persisted along stream beds, however. Makapansgat offers a similar picture of fairly extensive bush cover, but a drying trend has been detected at the site. There is still no archaeological record (no stone tools, in other words), and what we know both of the environment and of *A. africanus* itself suggests that no substantial changes had occurred in the early human way of life since the time of *A. afarensis*. Indeed, *A. africanus* remained so unspecialized many paleontologists think it may have given rise to all subsequent humans, although others prefer to see *A. afarensis* in this role.

Two crania attributed to *Australopithecus africanus*, both from the cave of Sterkfontein. On the right is the cranium Sts 5 (often known as "Mrs Ples," for the name *Plesianthropus* proposed for this form by Robert Broom), which, despite lacking teeth, has become the archetype of *A. africanus*. On the left is the partial cranium Sts 71, which may show some "robust" tendencies.

Photo by Willard Whitson.

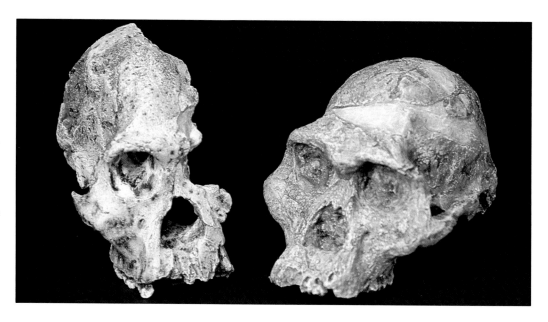

"Robust" early humans of southern Africa

In 1938, the first fossils of a new kind of early human began to turn up in South Africa at a place called Kromdraai, not far from Sterkfontein. These fossils revealed a hominoid whose jaws were more massively built than those of _A. africanus,_ and which had a flatter face. In the front of the mouth the incisors and canines were small, but behind them the chewing teeth were large and flat. Not only were the molars big, but the premolars in front of them were "molarized," providing a long, continuous grinding surface behind the canines. A modern human with molars this size would probably stand over ten feet tall! Large numbers of such "robust" specimens are now known from the nearby site of Swartkrans. Like the _A. africanus_ sites these places, neither far from Sterkfontein, represent rubble-filled solution cavities in the local dolomite and pose similar problems of dating.

A number of more or less complete crania (skulls without lower jaws) show that the chewing muscles were extremely large. Sometimes they were too large, in fact, to attach just on the surface of the braincase, with the result that in the biggest individuals a muscle-attachment ridge rises in the midline along the top of the braincase. In earlier times much was made of this gorilla-like feature, but actually it is there for purely mechanical reasons, and does not reveal any special relationship between the two hominoids. A large "sagittal" crest of this kind results simply from combining a small braincase with the large muscles needed to move a massive jaw. As in gorillas, though, possession of this crest helps to distinguish the larger males from the smaller females—though the size difference between the sexes is much smaller than in gorillas or in _A. afarensis._

These robust early humans were first described as members of a new species, _Paranthropus robustus_ ("robust next-to-man"); and although many paleoanthropologists have since preferred to classify them simply as another species of _Australopithecus,_ most are now returning to the view that indeed they should be placed in a separate but closely related genus. We will do so too, and refer to them as _P. robustus._ Despite their massive chewing apparatus, however, there is not much to suggest that the bodies of these early humans were appreciably bigger than those of _A. africanus_ (though their brains seem to have been slightly larger). Quite simply, we are still not sure how big these early humans actually were, since bones of the body skeleton of _P. robustus_ are rather few and far between. There are

View of the site of Sterkfontein, near Krugersdorp, South Africa. This site has yielded numerous early human fossils dating from over 2.5 million years ago to perhaps 1.6 million years ago.

Photo by Willard Whitson.

Palate and upper dentition of _Paranthropus robustus_ from Swartkrans, South Africa (Sk 13/14). The teeth of this "robust" form differ markedly from those of _Australopithecus_ species: the canines and incisors are reduced in size, while the premolar and molar teeth are greatly enlarged.

Illustration by Don McGranaghan.

Tooth wear and ancient diets. The chewing surfaces of teeth wear in various ways according to the different foods eaten. The very high magnifications obtained using scanning electron microscopes have allowed extremely detailed study of the wear left by chewing on the teeth of ancient humans. Here we see scanning electron micrographs at great magnification of the chewing surfaces of upper second molar teeth of *Australopithecus africanus* from Sterkfontein (above), and of *Paranthropus robustus* from Swartkrans (below). The micrographs reveal that the enamel is polished and scratched in *Australopithecus*, whereas in *Paranthropus* it is heavily pitted and gouged. This indicates that these two early humans chewed different kinds of foods, and that the diet of *Paranthropus* consisted of harder items than those eaten by *Australopithecus*.

Micrographs courtesy of Fred Grine.

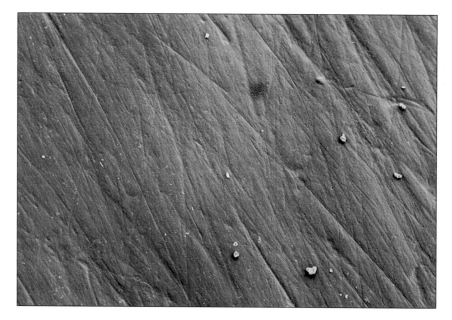

enough of them, however, for us to be certain that they walked erect like *A. africanus*. Indeed, the anatomist Randy Susman, of the State University of New York at Stony Brook, believes that foot bones from Swartkrans that may have belonged to *Paranthropus* show that its feet were more or less fully adapted to open-country bipedalism.

It's the important site of Swartkrans, too, that has provided the most direct evidence that some at least of the South African early human fossils were the victims of leopards. At this site the top of the skull of an immature *Paranthropus* was found that bore twin puncture-marks. Bob Brain, director of the Transvaal Museum and excavator of the Swartkrans site, precisely matched these holes with the size and spacing of the canine teeth of a leopard's lower jaw. According to Brain, this unfortunate young _Paranthropus_ must have been killed and dragged off by a leopard.

A different diet?

When researchers see differences as large as those that distinguish the highly specialized teeth of _P. robustus_ from the more generalized dentition of _A. africanus_, their immediate reaction is to conclude that the two had different diets. Thus as early as the mid-1950s the Transvaal Museum's John Robinson suggested that the "gracile" (lightly built) _A. africanus_ had a more omnivorous diet that included some meat (thus echoing Dart's ideas about a predatory origin for the human lineage). _Paranthropus robustus_, on the other hand, was committed to a more herbivorous, meatless way of life.

On the face of it, this is a very plausible conclusion. But detailed analyses of tooth wear undertaken since then have failed to support this dramatic difference in lifestyles. True, "robust" molars tend to be more pitted and chipped than "gracile" ones, but Fred Grine of the State University of New York at Stony Brook has argued cogently that this may mean nothing more than that the "robusts" habitually chewed on tougher, and perhaps smaller and grittier, objects than did the "graciles"; not that the latter ate more meat. "From the evidence of dental wear," Grine says, "it seems that both _Australopithecus_ and _Paranthropus_ fed principally on vegetation."

So what was the reason for this great difference? The answer probably lies in time and a changing climate. The "robust" sites of Swartkrans and Kromdraai pose the same kinds of problems of dating as do the "gracile" ones of Sterkfontein, Makapansgat, and Taung; but best estimates based on faunal comparisons would place them somewhere between about 1.9 million and 1.5 million years ago. They are, thus, substantially more recent; and in the period that separated the "gracile" and the "robust" sites the South African climate became much drier and the vegetation sparser. It is this shift to an open grassland environment, away from the more mixed habitat in which _A. africanus_ flourished, that seems to explain the specialized skull and dentition of _Paranthropus_. For they were well suited to chewing the tough, gritty foods—plant tubers and so forth—that are available out on the open grasslands, away from the forest.

Artist's reconstruction of a leopard dragging away the young Swartkrans _Paranthropus_ whose braincase was pierced by the canine teeth. Leopards frequently drag their prey into trees, which in the high veld often grow near the entrances to underground caverns. It is believed that at least some of the _Australopithecus_ and _Paranthropus_ bones from the South African sites accumulated in their underground burying places as a result of such activities.

Illustration by Diana Salles, after a sketch by Douglas Goode.

Two crania of *Paranthropus robustus* from the site of Swartkrans, about 1.6–1.9 million years old. They appear to be male and female, the female (on the left) being less robustly built and having a smaller attachment crest atop the skull for the temporalis muscle.

Photo by Peter Siegel.

As any ancestor must be, *A. africanus* was older and less specialized than *Paranthropus*. And, as we have seen, there is little to suggest that it could not have given rise to more specialized forms such as *Paranthropus* as it struggled to cope with a drying environment. Nonetheless, recent discoveries have made it clear that there were other early humans around to complicate the story.

The changing-habitat explanation for the differences between *P. robustus* and *A. africanus* may seem to lack drama, but perhaps it regains a little when we see how well the idea of habitat change fits with some recent finds at Swartkrans. In Member 1, the oldest part of that site, excavations over the past few years have produced examples of antelope horn cores (the bony protuberances on the skull over which the horns fit) and

View of the excavations at the South African cave site of Swartkrans, from which most of the known specimens of *Paranthropus robustus* have been recovered. An early form of *Homo* is also known from this site, as are stone tools and bones and horn cores that are polished from use as digging utensils. While the stone tools are generally considered to have been the work of *Homo*, many think it plausible that *Paranthropus* dug for roots and tubers with the polished utensils.

Photo by Willard Whitson.

Re-creation by the artist Jay H. Matternes of a scene near the South African site of Swartkrans, about 1.75 million years ago. A group of *Paranthropus robustus* disports itself on the open high veld. Some of the individuals are shown carrying horn cores or bones, or are digging with them. Such implements, polished by use, are known from the Swartkrans site; and while it is not certain that it was *Paranthropus* who wielded them, it is plausible that it was.

©1993, Jay H. Matternes; original painting in American Museum of Natural History.

broken bones that are undoubtedly polished by use. The excavator, Dr. Brain, set his sons to digging up roots and tubers with similar objects, and the result was a set of tools with more or less identical polishes. Rare bones of another form of human usually assigned to our own genus, *Homo*, are also present in Member 1, as are some crude stone tools.

Whether *Paranthropus* or *Homo* made the stone tools will continue to be debated, but *Paranthropus* may well have been responsible for the digging polish found on the horn cores and broken bones. Using such tools would have permitted access to buried tubers that are otherwise difficult to extricate, and chewing on these tough, gritty foods would very nicely explain the wear on *Paranthropus* teeth. Randy Susman also has stud-

The site at Olduvai Gorge, in northern Tanzania, where the first *Paranthropus boisei* fossil was found in 1959.

Photo by Willard Whitson.

ied hand bones from the same deposits that indicate a humanlike capacity for grasping, and if these bones belonged to *Paranthropus* (a minority view at present), they would strengthen the presumption that it was this early human that used the digging tools.

"Hyper-robusts" from East Africa

The first remains of an East African "robust" form came to light in 1959 at Olduvai Gorge, as the culmination of thirty years of laborious research there by Louis and Mary Leakey. Olduvai is a classic site on the plains of northern Tanzania where water action has worn away a deep cleft, thirty miles long and in places three hundred feet deep. In its walls are exposed sediments (identified from the bottom up as Beds I to V) that have accumulated over almost 2 million years. Olduvai is the first place at which the K/Ar dating method was used in paleoanthropology, and this new method proved, to the astonishment of all, how truly ancient the earliest humans represented there really were.

Equally astonishing was the size of this first "robust" cranium from East Africa, found near the bottom of Bed I in rocks just above a lava flow dated at around 1.8 million years old. In fact, it has been described as "hyper-robust," having a dentition of truly breathtaking massiveness. The front teeth are reduced even more than those of *Paranthropus robustus*; the premolars and molars are not just large but huge, providing an enormous flat grinding surface. Because of the diminutive front teeth its face is rather flat. With massive jaws and a brain volume of just 530 milliliters, this early human predictably has a large sagittal crest atop the skull. The new specimen was found in a level of the Olduvai sediments that produced stone tools. The new species (at first called *Zinjanthropus boisei*, but we'll call it *Paranthropus boisei* because it is not sufficiently different from the South African robusts to deserve its own genus) was at first acclaimed as the toolmaker, but the Leakeys soon transferred that honor to a primitive form of *Homo* whose fossils they found nearby a couple of years afterward. Who actually made the tools remains an open question, however.

More recently, through the efforts of the Leakeys' son Richard and an international group of colleagues, *P. boisei* has become very well known from fossil-bearing rocks that lie around Koobi Fora, east of Lake Turkana in northern Kenya. Its remains also have turned up at other sites in

The "Black Skull," KNM-WT 17000, discovered in sediments to the west of Lake Turkana in Kenya. Dated to about 2.6 million years ago, this specimen is believed by many to be near the common ancestry of both the southern and eastern African species of *Paranthropus*. With its very long face and its consequently elongated braincase, this fossil is often placed in the species *P. aethiopicus*, known otherwise only from very fragmentary material.

Photo by Willard Whitson.

Alternative names. As we've seen, paleontologists disagree about how many different species should be recognized in the human fossil record. The minimum number of species names that are generally accepted is listed in the left-hand column, while the one on the right shows to which of them the larger number of names accepted in this volume corresponds.

Australopithecus afarensis	*Australopithecus afarensis*
Australopithecus africanus	*Australopithecus africanus*
Australopithecus robustus	*Paranthropus robustus*
Australopithecus boisei	*Paranthropus boisei*
	Paranthropus aethiopicus
Homo habilis	*Homo habilis*
	Homo rudolfensis
Homo erectus	*Homo erectus*
	Homo ergaster
"Archaic" *Homo sapiens*	*Homo heidelbergensis*
Homo sapiens neanderthalensis	*Homo neanderthalensis*
Homo sapiens sapiens	*Homo sapiens*

Kenya and Tanzania, and in southern Ethiopia as well. These sites range in age from about 2.4 million to 1 million years, hence straddling the rather short period from which *Paranthropus* fossils are known in South Africa.

Paranthropus ecology in eastern Africa

During Bed I times the Olduvai basin was occupied by a lake, fed by streams that drained from highlands nearby. The human fossils are found in the lakeside sediments. Around the lake reed beds flourished, yielding away from the lake to trees and finally to a rather arid grassland. In the Omo basin of Ethiopia, *Paranthropus*-yielding deposits span a period in which the climate dried considerably; on the open plains vegetation became sparser with time, though forest presumably remained available along watercourses. A fluctuating

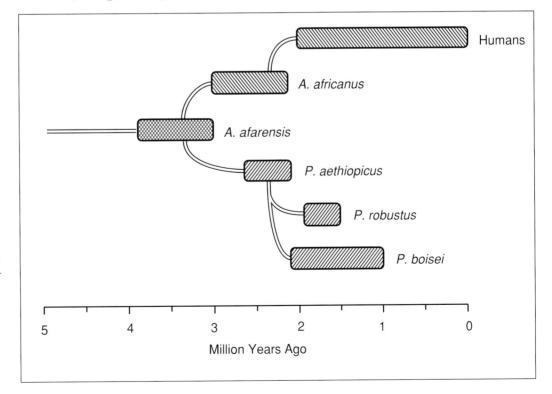

The most plausible scheme of evolutionary relationships among the species of *Australopithecus* and *Paranthropus*. The heavy bars represent the known time range of each species, and the lighter lines the proposed relationships among them.

Illustration by Diana Salles.

Comparison of two crania of *Paranthropus boisei* from east of Lake Turkana, Kenya. The specimen on the right is a male (KNM-ER 406); the much smaller and more delicately built specimen on the left (KNM-ER 732) represents a female. Both date to about 1.75 million years ago.

Photo by Willard Whitson.

environment also was typical of the time during which *Paranthropus* lived in the vicinity of Lake Turkana.

Unlike the situation in South Africa, then, the appearance of East African *Paranthropus* does not seem to have followed on the heels of a marked ecological deterioration. This may not be surprising, since a find made during the 1980s has pushed back the origin of "robusts" in East Africa by a considerable margin. This is the discovery to the west of Lake Turkana, in deposits about 2.6 million years old, of a rather complete skull (often known as the "Black Skull"; unfortunately it lacks teeth) that many paleoanthropologists assign to the species *Paranthropus aethiopicus*. The name is based on some rather scrappy specimens from the Omo basin that give the species a time range of 2.8–2.2 million years ago.

Relationships among the early bipeds

The complete skull of *P. aethiopicus* is clearly similar enough to other *Paranthropus* to be classified with them, but it is still distinctively different, most obviously in having a protruding and rather concave face. Oddly, it may have more in common with South African *P. robustus* than with its nearer neighbor East African *A. boisei*, but it also is claimed to retain primitive features otherwise present only in *A. afarensis*.

There has been quite a heated debate over the significance of this ancient form in early human evolution. At this point we have at least five players in the game, so there are lots of possibilities. The dust has yet to clear, but the simplest thing to conclude at the moment is that *A. afarensis* was a stem species that gave rise to at least two other species in the period around 3 million years ago. One of these was *A. africanus*, and the other was the ancestor of *P. aethiopicus*. In turn, *P. aethiopicus* quite rapidly spun off a lineage that split to become *A. boisei* in East Africa and *A. robustus* further south. But this scheme is probably greatly oversimplified (how many species don't we know about?) and therefore almost certainly incomplete, if not wrong. The only thing that does seem firmly established is that a little under 1 million years ago the "robust" forms became finally extinct, leaving no descendants behind them.

Chapter Seven

The Mystery of Olduvai Gorge: "Handy Man"

Olduvai Gorge is not only the site where the first East African *Paranthropus* was found. At the same time as *Paranthropus* lived around the ancient Olduvai lake another kind of early human also existed there, whose remains were first brought to light by the Leakeys in 1960. Nowadays it is generally accepted that it was these more advanced early people who made the simple stone tools that are found abundantly in the earlier deposits exposed in the walls of the gorge. But in 1964, when the fossil remains of this new kind of human were described as belonging to a member of a new, primitive member of our own genus, there was quite an outcry. The major complaint was that these fossils were not different enough from *Australopithecus africanus* to be in a species of their own, let alone a species of *Homo*, and it's certainly true that, presumed evidence of toolmaking apart, there was not a great deal on which to base this bold step.

The fossil bits and pieces in question consist of a slightly immature lower jaw, parts of the braincase of the same individual, and some bones of the body skeleton. These are from low in Bed I, and are about 1.8 million years old. These bones were named *Homo habilis* ("handy man") by Louis Leakey and two colleagues. At the same time these scientists associated with these finds a fragmentary braincase and a pair of upper and lower jaws that come from Bed II and are about 1.5-1.7 million years old. Later, a crushed but reconstructable 1.8-million-year-old skull from Bed I was added to the *H. habilis* roster, and quite recently a very fragmentary partial skeleton from low in Bed I has joined the ranks as well due to the efforts of Donald Johanson and his colleague Tim White.

The type lower jaw of *Homo habilis* (OH 7) from Olduvai Gorge, crushed and belonging to a young individual in whom the last molar was as yet unerupted. Associated with this specimen were two parietal bones and other skull fragments, from which a brain volume of 640 milliliters or more has been estimated. Some hand bones also were ascribed to the same individual.

Illustration by Don McGranaghan.

The first "advanced" humans

Apart from the evidence of stone toolmaking, the main difference from *A. africanus* that Leakey and his colleagues could point to among their *H. habilis* fossils was an apparent increase in brain size. The youthful braincase has been estimated at various volumes centering on 650 milliliters, and the fragmentary skull comes in at a fraction below 600 milliliters. These figures lie above the *A. africanus* range, although not by much. The Olduvai teeth, in contrast, are not very different from those of *A. africanus*.

As a result of this rather meager material, general acceptance of *H. habilis* had to await further discoveries. The most important of these were made during the 1970s in the Koobi Fora area to the east of Lake Turkana, and the find that finally tipped the balance in favor of *H. habilis* was the famous 1.9-million-year-old skull known as KNM-ER 1470, with its brain volume of about 750 milliliters. Ironically, more and more people are now coming to the conclusion that this skull represents not *H. habilis* but yet another early species of *Homo, Homo rudolfensis*. Nonetheless, other fossils from East Turkana, and some from the later deposits at South Africa's Sterkfontein (probably about 1.6 million years old and also associated with stone tools) do indeed appear to be comparable with the Olduvai specimens, and they help to vindicate *H. habilis* as a distinct and separate species.

Whether *habilis* should really be regarded as belonging to our own genus is debatable. The fragmentary Olduvai skeleton shows pretty clearly that this species was of very short stature, and that it retained the primitive short-legged body proportions of *Australopithecus* and *Paranthropus*. And some crania assigned to the species have notably small brains; one particularly good 1.9-million-year-old specimen from East Turkana has a braincase volume of barely over 500 milliliters. Still, habilis seems set to stay within *Homo* for the present, and as a relatively primitive form it remains a potential candidate for the ancestry of later *Homo*, although its late and rather restricted time span of about 2-1.5 million years ago makes this somewhat unlikely. For its part, *H. rudolfensis* may have been a bit more advanced in its body skeleton as well as in brain size, but it is nonetheless not generally seen as a direct human ancestor either.

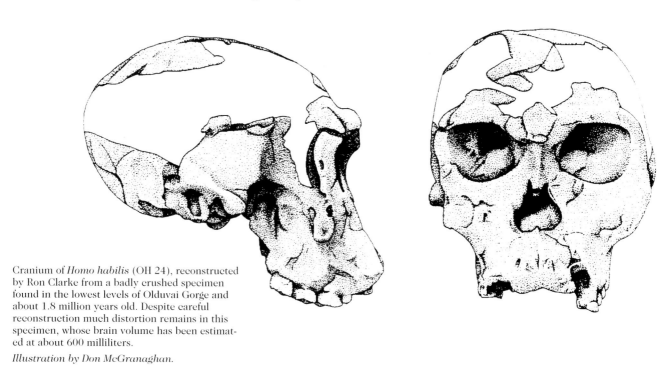

Cranium of *Homo habilis* (OH 24), reconstructed by Ron Clarke from a badly crushed specimen found in the lowest levels of Olduvai Gorge and about 1.8 million years old. Despite careful reconstruction much distortion remains in this specimen, whose brain volume has been estimated at about 600 milliliters.

Illustration by Don McGranaghan.

The fragmentary remains of the most recently discovered skeleton of *Homo habilis* from Olduvai Gorge, OH 62. The reconstructed limb proportions of this specimen are strikingly primitive, with relatively short legs and long arms—so much so, indeed, that Sigrid Hartwig-Scherer and Bob Martin of the University of Zurich have proposed that either the limb bones are wrongly attributed to *Homo habilis*, or that the assumed place of *Homo habilis* in human evolution should be revised.

Photo courtesy of the Institute of Human Origins.

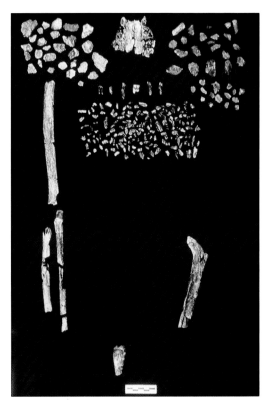

What is "culture"?

Whatever the relationships of these two species with later humans, however, it is with *H. habilis* and *H. rudolfens*is that we encounter the first uncontested association of early humans with stone tools. Stone tools have a particular importance to paleoanthropologists because their appearance is traditionally associated with that mysterious development, culture.

Anthropologists have debated endlessly over what "culture" actually is, mainly because it is really an arbitrary concept. Essentially what we mean by culture seems to be whatever we pass down by learning from generation to generation that our closest relatives, the apes, do not. And this boils down to a mishmash of symbolisms, beliefs and ideas about the world, not all of which have anything obviously to do with external reality. Where self-consciousness and language fit into the picture is not clear, although language is presumably necessary for the formulation and effective communication of abstract concepts, and self-consciousness is but a word that describes the "emergent" quality that gives us modern humans such a strong sense of being different from the rest of the living world.

The two most complete crania that have been attributed to *Homo habilis*. Both specimens come from east of Lake Turkana in Kenya, and are dated to about 1.9 million years ago. With its relatively large brain size of about 750 milliliters the specimen on the right, KNM-ER 1470, was instrumental in getting *Homo habilis* accepted as a distinct species. Ironically, however, it is now widely viewed as representing a separate species, *Homo rudolfensis*. The cranium on the left (KNM-ER 1813) is thought to belong to the same species as Olduvai *Homo habilis*, despite a smallish brain volume of about 510 milliliters.

Photo by Willard Whitson.

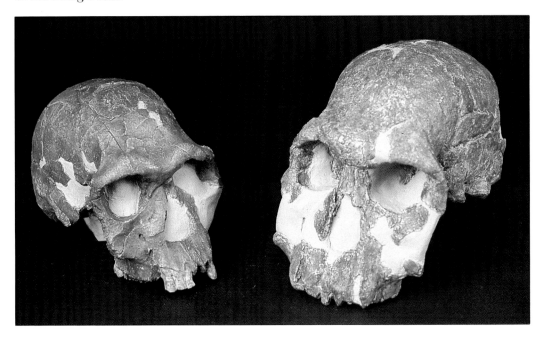

However this may be, making tools is not synonymous with what most of us would regard as "culture": after all, chimpanzees make rudimentary "tools." And it's certain that merely chipping stones to produce cutting implements doesn't equate with "culture" in anything like the sense in which we usually use the word, however poorly we may define it. Essentially, then, we're back here with the old problem of delineating "human." Nonetheless, it's clear that the production of stone tools was a fundamental step along the road to acquiring "humanity," opening up a vast range of new possibilities to a savanna-living biped and suggesting something to us about the reasoning capacities of the hominoids who first introduced it.

The first stone tools

The very first stone tools are considerably older than the presumed evidence at Swartkrans for the use by *Paranthropus* of horn cores and bones in digging up roots. Nonetheless, it does seem fair to suggest that the opportunistic use by our precursors of pieces of wood for digging may have preceded the flaking of stone—although there is obviously a limit to what early humans could have done with bits of wood without cutting surfaces to use to shape them. But inevitably such things must remain speculative in the absence of direct archaeological evidence.

Such evidence first appears at sites in eastern Africa that are dated to between 2.5–2 million years ago and that have yielded stone tools of so-called "Oldowan" type, though these sites have not produced evidence of the early humans who made them. The name Oldowan derives from Olduvai Gorge, where such tools were first identified, and which provides one of the best records of early human tool-using activities.

Oldowan tools consist of small pebbles of mostly volcanic rocks, usually about four or five inches long, from which one or two small flakes have been detached by blows from a hammerstone. Mary Leakey's pioneering early studies identified quite a variety of distinct tools among the Oldowan kit, but many archaeologists now think that it was the flakes, rather than the cores from which they were struck, that were most commonly used as implements. These flakes have sharp edges that were sometimes retouched with additional blows that removed smaller particles of rock and prolonged the cutting surface. Additionally, the experimental production of similar tools suggests that the different kinds of "core" tools that have been identified may have resulted from variations in

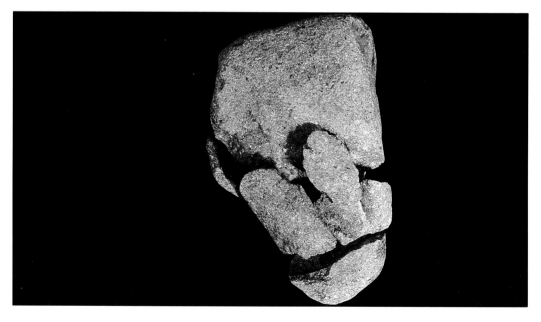

A replica of a stone tool of Oldowan type, consisting of a core, or small cobble, with several flakes knocked off it. Various kinds of cores have been identified, but at least as important in the tool kit were the small flakes with their sharp cutting edges.

Replica by Peter Jones; photo by Willard Whitson.

Stone circle discovered low in the sequence at Olduvai Gorge. Once interpreted as a windbreak, indicating some kind of "home base" of *Homo habilis*, it is now thought more probable that this circular arrangement of stones is the result of natural forces.

*Photo by
Willard Whitson.*

the pebbles used rather than from an intention on the part of their makers to produce tools of specific types. Different materials are available in different areas, and it seems that the toolmakers made a careful choice among the pebbles available to them, sometimes carrying them some distance to the point at which they were eventually found by archaeologists.

The production and use of experimental stone tools also has permitted archaeologists to study the various kinds of wear on cutting surfaces that different uses produce. Analysis of an Oldowan-type tool assemblage from Koobi Fora, for instance, showed that some were used to scrape or to saw wood, some for cutting grass or reeds, and others for cutting meat. Experiments also have shown that the larger cores were probably useful in the less-delicate aspects of butchering. Still, many of them probably were simply the byproducts of flake production.

Of course, archaeologists are not concerned only with individual artifacts. Where artifacts are associated in the sediments with animal bones (some of which do carry distinctive "cut marks" made by stone tools) and unusual arrangements of natural objects, they also try to understand what these associations tell us about early human activity. Caution is in order, though. During early work at Olduvai a locality was discovered in Bed I—near where *H. habilis* fossils had been found—that contained animal bones and an unusual circular arrangement of stones. This was first identified as a *H. habilis* living site whose early human inhabitants had butchered prey and built a windbreak for shelter. But later work has shown that the stone ring probably resulted from fracturing of the underlying lava by the roots of a tree, and that the animal bones were deposited by stream action at the places where they were found. This echoes work at Koobi Fora and elsewhere that suggests that we cannot always make too much of the association of artifacts and animal bones, especially at lakeside sites where water action was prevalent.

Cautious research of this kind has reduced considerably the drama of the picture that we can draw of the lives of our precursors—and not just at Olduvai. Back in the

1970s it was popular to interpret sites such as the one at Olduvai as "home bases," where ancient humans repeatedly returned to shelter and to carry out various activities including the butchering of prey. This, it was pointed out by the distinguished archaeologist Glynn Isaac among others, is a behavioral trait that strongly distinguishes modern humans from living apes, and it was felt to be a fundamental aspect of the behavior patterns that differentiate the one from the other. By extension, it also was taken to imply that by about 2 million years ago other distinct human activities had arrived on the scene—for example, food sharing and the division of labor between the sexes.

Now archaeologists are less confident about this interpretation. They also are less willing to draw close analogies between modern hunting-gathering peoples and the behavior of early humans who also spent their lives on the move. Meat forms a substantial (if generally overestimated) part of the diet of modern hunter-gatherers, but it is by no means obvious that this was necessarily true of the earliest tool-using humans.

The lifestyle of *Homo habilis*

Despite the fact that it made tools, the lifeways of *H. habilis* probably did not differ too dramatically from those of *Australopithecus* and *Paranthropus*. It remained small-bodied and with primitive limb proportions. It existed in a habitat in which forest was becoming scarcer, and through which it presumably traveled pretty constantly in bands large enough to provide protection in the open while still remaining small enough to be supported by the relatively unproductive (for a hominoid) savanna environment. Its brain remained rather small, and its means of communication were almost certainly unsophisticated.

Tool use would certainly have allowed *H. habilis* to scavenge medium-sized and larger mammal remains more efficiently than *Australopithecus* and its contemporary *Paranthropus* were able to, but it is unlikely that it actually hunted anything larger than the smallest mammals. Most paleoanthropologists would guess that vegetable food items (which do not preserve in archaeological sites) furnished the greater part of its diet, and that they were supplemented by small vertebrates, insects, and so forth. Beyond this it's impossible to infer anything with much confidence, though techniques are advancing so rapidly that archaeologists may soon be able to resolve many of the ambiguities that currently plague them. One thing, though, does seem clear: the great leap forward in human evolution was yet to come.

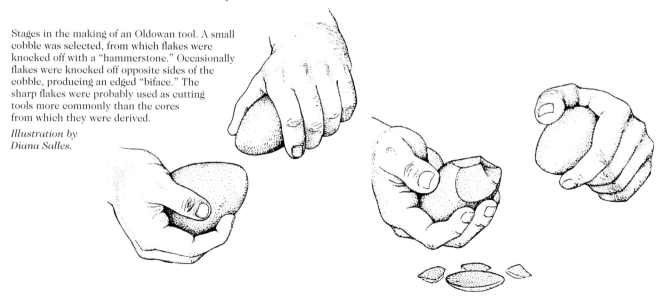

Stages in the making of an Oldowan tool. A small cobble was selected, from which flakes were knocked off with a "hammerstone." Occasionally flakes were knocked off opposite sides of the cobble, producing an edged "biface." The sharp flakes were probably used as cutting tools more commonly than the cores from which they were derived.

Illustration by Diana Salles.

Chapter Eight

The Great Leap Forward

Before *Homo habilis* was described in 1964, the earliest known member of our own genus *Homo* was *Homo erectus*. The discovery of this early human more than seventy years earlier makes up one of the most extraordinary stories of inspiration combined with perspiration in the entire annals of paleontology.

The young Eugene Dubois, as he appeared shortly before his departure for the East Indies.

Portrait drawn by Diana Salles from a photograph.

The "upright ape-man" from Java

In 1887 Charles Darwin's *On the Origin of Species* had been in print for less than thirty years, and the only human fossils known were a handful of large-brained and rather recent specimens which scientists of the time had a lot of trouble interpreting. To all intents and purposes, the science of paleoanthropology had yet to be born. So nobody fully understands why in that year a young Dutch anatomist named Eugene Dubois (1858-1940) quit his job at the University of Amsterdam, enlisted as a military doctor in the Royal Dutch East Indies Army, and sailed to what is now Indonesia with the avowed aim of finding the remains of fossil man. This was not just like looking for a needle in a haystack, for it was a pure gamble that the "needle" was there to be found at all in that vast 3,500-mile-long island chain. Dubois himself never recorded exactly why he took this plunge. But maybe he was motivated by Darwin's prophesy that the ancestors of humankind would be found in Africa, the home of two out of the three great apes. For while the Dutch had no colonies in Africa, they did then own Indonesia, the home of the third great ape, the orangutan.

Whatever Dubois's reasons for giving up a promising academic career to become an army surgeon and part-time fossil collector, his gamble paid off handsomely, if not right away. His initial explorations on the island of Sumatra produced only one fossil human skull, and this was clearly a relatively recent example of *Homo sapiens*, modern man. But in 1890 Dubois transferred his excavations to the neighboring island of Java, where the colonial authorities provided him with a large gang of convict laborers. In late 1891 this crew found a distinctly humanlike skullcap at the bottom of a huge pit dug on the banks of the Solo River, near a village called Trinil. And not long afterward a single human thigh bone was found not far away. Eventually Dubois amassed several other bits of thigh bone, plus a tooth which was the first of the Trinil human fossils to be discovered.

The odds against such a discovery in this island terra incognita were enormous, and Dubois's finds caused quite a stir when news of them got back to Europe. However, Dubois had considerable difficulty in getting

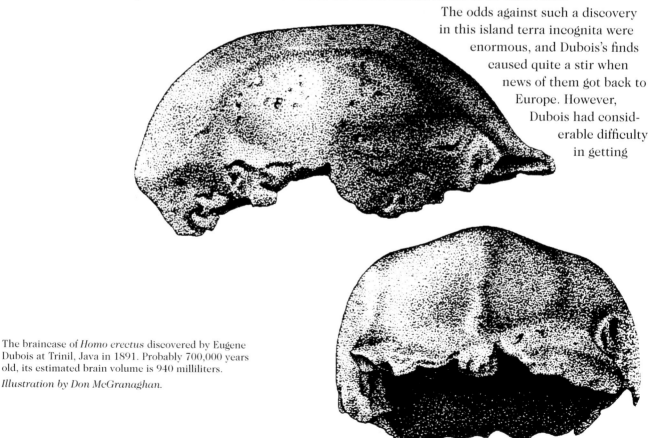

The braincase of *Homo erectus* discovered by Eugene Dubois at Trinil, Java in 1891. Probably 700,000 years old, its estimated brain volume is 940 milliliters.

Illustration by Don McGranaghan.

Photograph of Eugene Dubois's excavations at Trinil, on the banks of the Solo River, in Java, taken in 1894.

Courtesy of John de Vos/Nationaal Naturhistorisch Museum, Leiden.

them recognized as human precursors, partly because some of them, at least, were so different from anything else then known. The complete thigh bone was, except for a bony outgrowth, very similar to the same bone of modern humans—so similar, in fact, that doubts were raised as to whether the two fossils really belonged to the same species. But the other find really raised the eyebrows of the scientific world.

This was the truly unusual skullcap: long and low, with a sharply angled rear and with distinct bony ridges overhanging the eyes in front, rather like a gorilla or a chimpanzee. The braincase was also thick-walled, and it had contained a brain of only about 940 milliliters in volume. As we know today, this is significantly larger than the brain of *H. habilis*; but it is still a long way short of the modern average of 1,350 milliliters. In choosing a name that reflected both the primitive skull shape and the modern human posture indicated by the thigh bone, Dubois dubbed his new species *Pithecanthropus erectus* ("upright ape-man").

The best-preserved of the *Homo erectus* crania known from central Java (Sangiran 17). Discovered in 1969 and thought to be about 800,000 years old, this fossil, possibly of a male, is more masssively constructed than specimens collected earlier. Brain volume is about 1,000 milliliters.

Photo by Willard Whitson.

Although today we know with reasonable confidence that Dubois's fossils from Java, and others discovered since, date from about 1 million years to 700,000 years ago, early doubts about

"Java Man" were also encouraged by uncertainty as to their age. Dubois reckoned that the fossil mammals found in the same deposits had lived in the late Pliocene epoch or the early Pleistocene (which would have made them, we know today, about 1.5–2 million years old). But since at the time there were few well-dated fossil mammals known from Asia to compare them with, this was a rough estimate at best, and it left lots of room for argument.

Moreover, after 1912, interpretation was yet further complicated by the "discovery" in England of the famous Piltdown Man. This fraudulent "specimen" was thought to be older than the Java fossil but it had a much larger braincase. If the large human brain had been the first human feature to emerge, as Piltdown seemed to show, then the later but smaller-brained Java Man looked like an anomaly. And finally, we should remember that before Dubois's discoveries the only fossil humans known were the relatively recent Neanderthals of Europe, and these had brains fully as large as our own. So a hominoid that combined a smallish brain with upright stature was quite outside the experience of even the most seasoned scientists of the time.

Today Piltdown is a fading memory, and all scientists accept that Java Man represents an early human species. Indeed, for the past four decades this species has been placed in our own genus *Homo*, as *H. erectus*. But for the first part of this century the human status of Dubois's fossils had to await confirmation. Some of the new fossils which provided this confirmation came from Java itself, but probably the most influential discoveries were made at the Chinese site of Zhoukoudian, near Beijing, in the 1920s and 1930s.

Peking Man

In China the fossils of extinct animals have for millennia been revered as "dragon bones," with wonderful medicinal properties. Such fossils are thus eagerly sought by the proprietors of traditional drug stores. One of the sites from which such objects came was an abandoned lime mine at Chou K'ou Tien (now Zhoukoudian), not far from Beijing. Around 1920 prospecting at this site was begun by a Swedish mining engineer, Johan Gunnar Andersson (1874-1960). A mining engineer by profession but a paleontologist by avocation, Andersson had acquired a paleontological monopoly from the Chinese government, and by 1921 he had found not only mammal fossils but crude stone tools at Zhoukoudian. The discovery of two human teeth there fired the interest of Davidson Black (1884-1934), professor of anatomy at the Peiping Union Medical College, and in 1927 major excavations were begun by the college under the direction of a young Swedish paleontologist named Birger Bohlin (b. 1898). By the time excavations were suspended ten years later because of increasing guerilla activity in the area, the broken remains of more than forty individuals, initially given the name of *Sinanthropus pekinensis* ("Chinese man of Peking"), had been recovered from Zhoukoudian.

From the moment the first braincase of "Peking Man" was discovered in 1929 by the Chinese paleontologist W.C. Pei (b. 1904), it was evident that here was an early human very much like the one from Java. Five fairly complete braincases from Zhoukoudian had contained brains ranging from about 850 to 1,200 milliliters in volume, and they all showed the same basic features as Java Man, with long, low skulls and receding foreheads. As a result of these similarities the rather rapid acceptance of the Chinese fossils as human meant welcoming the Java ones into the fold as well.

Finally, Dubois was vindicated—although by this time, perhaps at least partly in reaction to the rejection of his own interpretations over four decades, the aged Dubois himself was proclaiming the remains to be those of a giant extinct gibbon! As for the

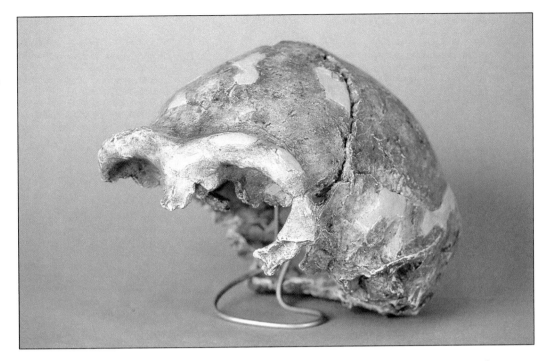

The most complete of the "Peking Man" brain-cases found at the site of Zhoukoudian, near Beijing, this specimen (CKT Skull III) preserves a small part of the face. Probably 300,000–400,000 years old, its brain volume is about 1,100 milliliters.

Photo by Willard Whitson.

Beijing fossils, though, Dubois considered them to be "perfectly human." Dating them is tricky, but it's now generally reckoned that the cave deposits from which they came accumulated over a long period, between 600,000 and 200,000 years ago.

Cannibal feasts?

It was the detailed anatomical studies carried out by Franz Weidenreich (1873-1948), Davidson Black's successor as professor of anatomy at Peiping Union, that showed beyond doubt that Peking Man lay in the human lineage. But this conclusion was certainly also helped along by archaeological evidence.

Like so many other human fossil localities, the Zhoukoudian site represented a cave that had been formed by solution in limestone, and in 1931, layers of carbon found in the rubble within were interpreted as places where fires had burned. Crude quartz tools also found at the site endowed Peking Man with stoneworking technology as well as the control of fire. And the picture of Peking Man took on a yet more dramatic aspect when Weidenreich suggested that the human remains found in the cave were the remains of cannibal feasts. Weidenreich was convinced that this was why all of them were broken, and why not one complete human skeleton had been found at Zhoukoudian. "The *Sinanthropus* population of Choukoutien," he wrote, "had been slain, and . . . subsequently their heads were severed from the trunk, the brain removed and the limbs dissected."

Later archaeologists have rejected this story, favoring a more prosaic explanation of breakage by hardworking scavengers and other natural agents. But the idea of cannibalism as well as the use of tools and fire certainly helped to imbue Peking Man with a human image if a savage one, and by the time the fossils were lost in 1941 during the confusion created by the Japanese invasion of Beijing, the position of Peking Man and Java Man as linear human ancestors seemed secure. So secure, indeed, that in the early 1950s the name *Sinanthropus*, like *Pithecanthropus*, disappeared from the literature of paleoanthropology at the urging of the distinguished evolutionary biologist Ernst Mayr (b. 1904), and Peking Man joined Java Man as a member of the species *H. erectus*.

"Homo erectus *Utilizing a Prairie Fire,*" painting by Jay H. Matternes, set near the Zhoukoudian site about 400,000 years ago.
From Peoples and Places of the Past, published by the National Geographic Society, © 1982, Jay H. Matternes.

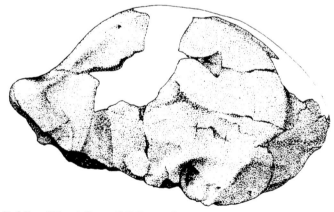

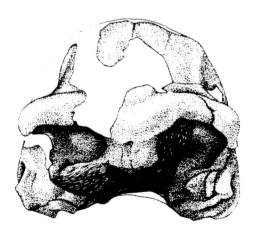

Braincase from Bed II at Olduvai Gorge (OH 9), usually regarded as belonging to *Homo erectus*. About 1.2 million years old, this specimen is more robustly built than the fossils of *Homo ergaster*, known from half a million years earlier at East Turkana, and the typical *Homo erectus* specimens from Java. If its affinities are with *Homo erectus* as typified by the Sangiran 17 specimen, this specimen may indicate that *Homo erectus* evolved in Africa before spreading to other parts of the world. Brain volume is 1,067 milliliters; scale is 1 centimeter.

Illustration by Don McGranaghan.

Homo erectus in Africa?

By the mid-1950s, then, Java Man and Peking Man were accepted by almost all anthropologists as direct ancestors of *H. sapiens*. At the same time the Piltdown fraud was uncovered, taking Europe out of contention as the place where early *Homo* had evolved. Eastern Asia thus enjoyed the limelight as the uncontested focus of human evolution during the middle part of the Pleistocene epoch (the period which lasted from 1.6 million years ago almost up to the present). Africa, despite its eminence as the continent with the most primitive humans—*Australopithecus* and *Paranthropus*—had little to offer that was comparable to *H. erectus*. Some lower jaws found at Tighenif in Morocco in the 1950s were thought to resemble *H. erectus*, and in 1960 a heavily built 800,000-year-old skullcap with a brain volume of 1,067 milliliters was found in Bed II at Olduvai. But that was about it.

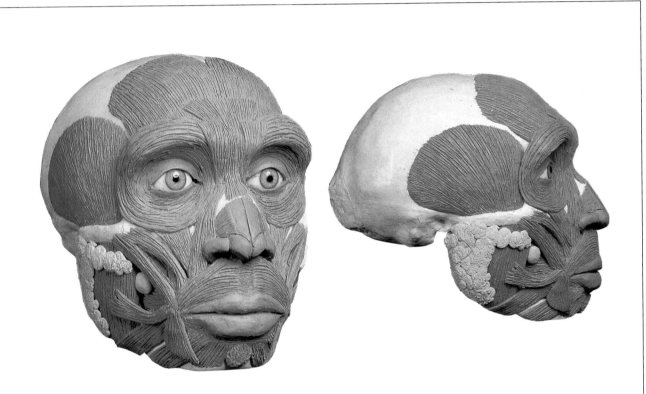

Reconstructing an extinct human. There is nothing like a three-dimensional flesh reconstruction to bring the past alive in the mind's eye, so several reconstructions of extinct humans were made for the American Museum of Natural History's *Hall of Human Biology and Evolution* for display in diorama settings. This is not a simple process. Here we see a stage in such a reconstruction, this one of an aged female Neanderthal. From an incomplete skull, museum technician Gary Sawyer first made a complete skull reconstruction. Using modeling clay, he then added the muscles, layer by layer, building up from the deep muscles close to the bone, to the superficial muscles, and finally to the fat deposits and connective tissues that underlie the skin. The photograph shows how the reconstruction looked just before the skin and associated tissues were finally added. Although we are thus assured of the basic accuracy of the reconstruction, many of the external details (including important ones such as ear, nose and lip shape) remain a matter of judgment. Features added at the end, such as the color, texture, and distribution of the hair, and the color of the eyes and skin, are also purely conjectural. Nonetheless, the painstaking process we see here ensures as accurate a depiction as possible of basic Neanderthal appearance.

All this was dramatically changed during the 1970s and 1980s by a team led by Richard Leakey, then director of the National Museums of Kenya. During this period Leakey and his colleagues made a remarkable series of finds on both the eastern and western shores of Lake Turkana. Among other fine specimens is a relatively complete adult cranium from East Turkana (known by its museum number of ER 3733) that is about 1.7 million years old, with a skull volume of about 850 milliliters. This specimen looks quite similar to *H. erectus* but its braincase is more highly arched and more lightly built, and its face is less massively constructed than in the few *H. erectus* specimens that preserve this feature. There are also various other minor differences.

Several other fossils from the Turkana Basin have been attributed to the same species, but among them ER 3733 is surpassed only by a slightly younger and amazingly complete skeleton of a youth from West Turkana. Discovered by the legendary Kenyan fossil collector Kamoya Kimeu, this ancient and remarkably complete skeleton is a truly extraordinary find.

The "Turkana Boy"

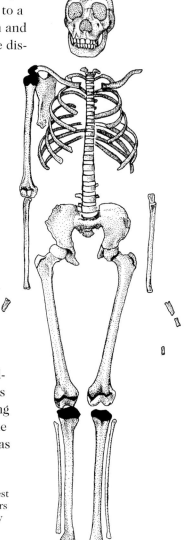

The skeleton in question is that of a young male who died at perhaps nine years of age, at a developmental stage equivalent to a modern eleven-to thirteen-year-old. He perished in a marsh and was rapidly covered over by mud before his corpse could be dismembered by scavengers, which is why his remains are so astonishingly complete. At death this boy stood about five feet, three inches tall and weighed about 106 pounds. However, the paleoanthropologist Alan Walker has estimated that had he lived to adulthood he would have grown to a height of six feet, one inch, considerably above the modern average, and would have weighed about 150 pounds. Moreover, certain details aside (such as a chest that tapered rather sharply upward, as did Lucy's), the body proportions of the "Turkana Boy" were perfectly modern: "He was built," says Walker, "just like the modern inhabitants of the Lake Turkana region: not only tall but slender, with long limbs."

These body proportions, Walker thinks, reflect the need, then as now, to shed body heat in the hot, dry, open Turkana environment: they increase the body's heat-losing surface area relative to its heat-generating volume. Such specimens as the thigh bone discovered in Java by Dubois had long suggested that *H. erectus* had achieved totally modern upright walking perhaps 700,000 years ago; but here was proof that by about 1.6 million years ago people were walking around in Africa with bodies that had major features just like our own. They thus differed radically from such precursors as *H. habilis* from Olduvai Gorge, a mere 200,000 years older!

Drawing of the "Turkana Boy" skeleton (KNM-WT 15000), discovered in 1984 at Nariokotome, to the west of Lake Turkana, Kenya. This remarkably preserved skeleton is that of a youth who died 1.6 million years ago and who, it is estimated, would have grown to over six feet tall had he survived to maturity. His body proportions are strikingly modern, with long, slender limbs, although his chest cavity tapered inward rather sharply toward the top. His species would later be identified as *Homo ergaster*.

Illustration by Diana Salles.

The famous KNM-ER 3733 cranium from East Turkana, Kenya, the best-preserved skull specimen of *Homo ergaster*. About 1.7 million years old, this fossil has a braincase volume of about 850 milliliters.

Photo by Willard Whitson.

Advanced toolmaking: hand axes and cleavers

These finds demonstrate that by about 1.8 million or 1.9 million years ago (the two fossils described above are not quite the earliest of their kind), early advanced humans with smallish brains but bodies very much like our own lived alongside *H. habilis*, *H. rudolfensis*, and *Paranthropus boisei* in the Turkana region. By this date there was already a flourishing stoneworking industry at East Turkana, making tools that don't differ much from those found at the bottom of Olduvai Gorge. Any of these early humans might have made these tools, and presumably all of the species of *Homo* did.

However, about 1.5 million years ago, as *H. habilis* and *H. rudolfensis* died out, East Turkana and other African sites began to produce the more complex tools usually known as "Acheulean," for St. Acheul in France where such implements were first identified in the 1830s. Measuring six inches to nearly a foot long and carefully shaped on both sides—hand axes to a point at one end, cleavers to a broad edge—these are clearly tools that

Geologist Craig Feibel prospects for fossils at East Turkana, near the spot where the cranium KNM-ER 3733 was found (indicated by the concrete marker).

Photo by Willard Whitson.

A pointed hand axe and blunt-ended cleaver from deposits at St. Acheul, France. Both of these implements are typical of the "Acheulean" tool-making tradition, which derives its name from this site.

Photo by Willard Whitson.

were formed to a pattern that existed in their maker's mind. Older ways were not immediately abandoned, though: the simpler tools of the Oldowan tradition continued to be produced alongside the hand axes. Indeed, in parts of eastern Asia these cruder implements continued to be the dominant tool types until not many thousand years ago, and hand axes are rarely found. Bamboo is common in these regions, and Geoffrey Pope of William Patterson College has suggested that tools made from this giant grass might have fulfilled many of the cutting and scraping functions that were performed by stone elsewhere.

The appearance in the Turkana area of a new kind of human a long time before the introduction of new kinds of tools illustrates a pattern that holds throughout the length of the Paleolithic period (otherwise known as the "Old Stone Age"). This is that new stoneworking technologies never seem to begin with a new species of human. It seems counterintuitive, for wouldn't a new, improved human be the best explanation for a new, improved technology? Actually, no: innovations must arise within species, for there is no place else that they can do so. This is true whether such novelties are technological (such as a new kind of tool) or anatomical (upright posture, for example). And, as we saw earlier, this discrepancy leads to problems when you try to group fossils into species in the fossil record.

Acheulean hand axes, cleavers and other stone implements litter the ground at Olorgesailie, an 800,000-year-old site in central Kenya.

Photo by Willard Whitson.

How many species?

When "advanced" humans were discovered in the Turkana Basin in sediments close to 2 million years old, scientists were impressed by their resemblance to *H. erectus*, although Richard Leakey and his colleagues cautiously described them simply as *"Homo sp."* (unidentified species of *Homo*). Today it is fairly common to call them *H. erectus*. But are they *H. erectus*? I think not. Nor do I think that *H. erectus* is our direct ancestor. Here's why.

In 1975, the same year in which ER 3733 was discovered, my colleague Niles Eldredge and I pointed out that *H. erectus* possessed a number of specialized features that later humans don't have. In particular, the very long, low skull and the thickness of the bone composing it make *H. erectus* an unconvincing intermediate between *Australopithecus africanus* and *H. sapiens*, both of which have relatively light, vaulted braincases. As long as *H. erectus* remained the only known candidate for the ancestry of *H. sapiens*, people were prepared to overlook this inconvenient fact. Once ER 3733 was described from Turkana, however, it became evident that here was a much more plausible ancestor for our own species. Its skull shape retains the lighter, more vaulted build that one would expect in a human precursor. Thus many scientists now believe that this early advanced human from the Turkana Basin (and maybe others like it from South Africa) was the ancestor of both *H. erectus* on the one hand, and of the lineage leading to modern humans on the other. *Homo erectus* thus becomes an offshoot from the human lineage, a side branch that may have evolved its specialized skull features in the isolation of eastern Asia.

If the Turkana species was not *H. erectus*, what should we call it? In 1975 Colin Groves of the Australian National University and his colleague Vratislav Mazak of the National Museum in Prague, then Czechoslovakia, applied the new species name *H. ergaster* ("Work Man," in acknowledgement of its toolmaking) to a lower jaw that had been found at East Turkana several years before ER 3733 turned up. Lower jaws don't have a lot of features that are useful in defining species, but most scientists now consider that this specimen belonged to the same species as the cranium ER 3733. If this association is correct, then the Turkana species should be called *H. ergaster*, and this is the name we will use here.

Lifestyles of early advanced humans

Despite the highly dramatic scenarios that were painted of the lifeways of *H. erectus* and its relatives in earlier times (driving mammoths into swamps to mire and butcher them, for example), most archaeologists today are cautious in what they say about the lives of *H. erectus* and *H. ergaster*. They are particularly reluctant to conclude that these early humans systematically hunted large animals. It is still difficult at most sites to determine if animal bones are the work of nonhuman scavengers, of human hunters, or of human scavengers. Further, it is probable that complex hunting techniques were not developed until much later times, perhaps only after *H. sapiens* came on the scene.

Yet *H. ergaster* was probably the first human to domesticate fire. At Chesowanja and perhaps at other sites in Kenya of roughly equivalent age, burned balls or patches of clay suggest that fire was used about 1.4 million years ago; and at a slightly earlier time there also is evidence of fire use at South Africa's Swartkrans, where stone tools have been associated with the remains of an early human similar to East African *H. ergaster*. Both of these instances remain controversial, however, and it is only at much later sites such as Zhoukoudian and various European localities that we can be fully confident that fire had finally been domesticated by humans.

Sites yielding fossils belonging to the *Homo ergaster/Homo erectus* group.

Illustration by Diana Salles.

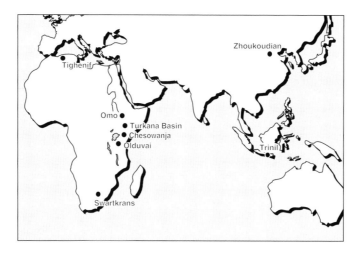

What we can be certain of is that by about 1.7 million years ago people who looked pretty much as we do below the neck were living out on the African savannas. These were people who were definitely committed to an open-country existence, and who had become tall enough to decrease some of the dangers of savanna life. During their tenure on Earth they introduced a new way of making tools: one that clearly suggests that the toolmakers, despite their relatively small brain sizes, were manufacturing implements to a standard pattern. By scaring off predators their control of fire would have increased their security in the open environment, even if it was not essential for the way of life that had governed their physical shift in the modern direction. To modern people fire has a great deal of symbolic meaning, including a sense of domestication and of the "otherness" of the world beyond the ring of flickering light; whether it came to have the same meaning for the first fire users, we'll probably never know.

However, we can say rather little about the mental capacities of these people, although we can be fairly confident that spoken language as we understand it today lay well in the future. It's likely, however, that gestural and vocal communication between individuals was considerably more complex than it was among *H. habilis*. If we can fairly say that a "great leap forward" occurred in human evolution at any point before modern humankind came along, it took place in that obscure gap between *H. habilis* and *H. ergaster*.

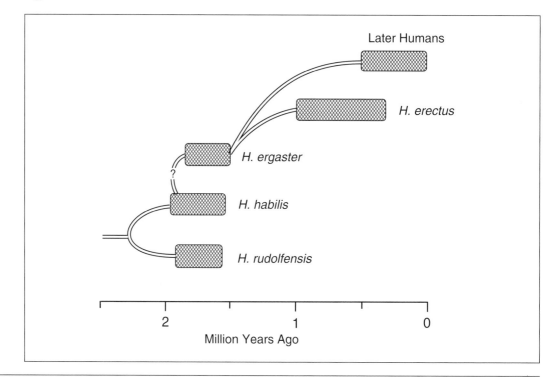

The evolutionary relationships among the various early members of the genus *Homo*. The dark bars show the known time span of each species in the fossil record; the lines indicate possible relationships among them.

Illustrated by Diana Salles.

Chapter Nine

Toward Modern Humans

Among primates the emergence of new species demands that old species be fragmented into isolated populations, in which the necessary genetic changes can take place. The needed environmental disruptions occur most frequently during periods when climates are unusually unstable, and at no time have climates fluctuated more aggressively than during the Pleistocene epoch, which lasted from about 1.6 million years ago nearly up to the present. Almost certainly, the accelerating pace of change in human evolution over the course of the Pleistocene was due to these special conditions. Let's look at what caused them.

The Ice Ages

The Pleistocene has earned the epithet "Ice Ages" because during this time the climate swung wildly between temperate periods much like today, and colder ones marked by major ice buildups in the northern hemisphere. Such extreme changes had not occurred for a very long while. They resulted from the combination of normally harmless shifts in such things as the tilt of the Earth's axis, the shape of its orbit, and the amount of atmospheric carbon dioxide.

Following a "cold snap" about 2.5 million years ago, the late Pliocene was a relatively warm time, during which the average global temperature was never lower than today's.

As the ice sheets enlarged during cold periods, "locking up" water on land, sea levels fell. This map shows how western Europe would have looked during glacial maxima. Most of the British Isles was covered by thick sheets of ice (white areas), as were mountainous areas of continental Europe; the dark areas show how Britain was connected to the continent by a land bridge exposed by retreating seas.
Illustrated by Diana Salles.

Atlantic Ocean

Mediterranean Sea

The next major climatic cooling took place about 1.6 million years ago, as the Pleistocene began, and was again followed by milder times. Cold conditions then returned about a million years ago, and the climate settled down to a rhythm where global temperatures went from one cold peak ("glacial") to the next at approximately 100,000-year intervals, passing through a short warmer phase ("interglacial") between the peaks. The last glacial ended about 11,500 years ago, and the time since then is sometimes referred to as the Holocene epoch. But there is in fact little justification for using this special name since there is no indication that we have yet emerged from the Pleistocene climatic cycle.

The Ice Ages. This diagram shows how temperatures have oscillated over the course of the Ice Ages that marked the Pleistocene epoch, and which so profoundly affected the course of human evolution. The Pleistocene was not a time of simple cooling; rather, global temperatures fluctuated in a complex fashion. This is shown by the curve on the right, extrapolated from the relative abundance in marine fossils of heavy and light oxygen isotopes. Heavy isotopes are more common in colder times. The vertical black-and-white bar represents the condition of the Earth's magnetic field, which periodically reverses itself. Since this field imprints itself on certain kinds of rock as they form, it has provided a useful tool in geological dating. The black segments represent periods during which magnetism was "normal," like today; the white ones represent "reversed" periods, when compasses would have pointed toward a southern magnetic pole. Major events in human evolution are listed along the time scale.

Illustration by Diana Salles.

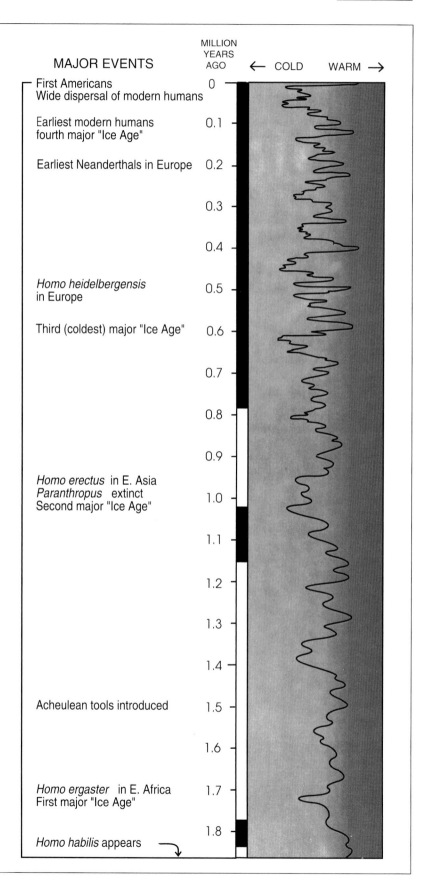

MAJOR EVENTS	MILLION YEARS AGO
First Americans Wide dispersal of modern humans	0
Earliest modern humans fourth major "Ice Age"	0.1
Earliest Neanderthals in Europe	0.2
	0.3
	0.4
Homo heidelbergensis in Europe	0.5
Third (coldest) major "Ice Age"	0.6
	0.7
	0.8
	0.9
Homo erectus in E. Asia *Paranthropus* extinct Second major "Ice Age"	1.0
	1.1
	1.2
	1.3
	1.4
Acheulean tools introduced	1.5
	1.6
Homo ergaster in E. Africa First major "Ice Age"	1.7
Homo habilis appears	1.8

COLD ← → WARM

Glaciation is the name given to the buildup of ice on the earth's surface, and to the effects that such ice has on the underlying landscape. Ice doesn't just sit there; it moves, and it has a profound effect on the earth over which it moves. Even today, in an interglacial period, the Earth boasts ice caps at each pole, and when we speak of Pleistocene glaciation we refer to the periodic expansion of those ice caps. These expand when the temperature drops because snowfall builds up over much larger areas. At such times, too, the snowcaps of mountain ranges expand to cover large parts of the surrounding landscape, sometimes melding with the expanding polar ice caps and in other places forming huge barriers to human movement.

During parts of the Ice Ages, ice buildups covered vast areas of what are today some of the most densely inhabited regions of the world. For instance, about 20,000 years ago, at the height of the last glaciation, most of the north-central United States was covered by an ice sheet two miles thick. Lobes of this sheet reached as far south as Kansas City and New York, where Long Island was formed as the retreating glaciers dumped the enormous load of rubble that they had picked up on their journey south. Interestingly, this ice sheet was centered not on the North Pole but on a point to the west of Hudson Bay. In Europe an ice sheet centered on Sweden buried London and points north, and the ice cap of the Alps formed an enormous barrier between the Mediterranean and central Europe. Areas not directly invaded by the ice caps were, of course, affected by their presence, and changing global climates affected environments worldwide, although not in the same way everywhere.

The seas are the ultimate reservoir of the world's water supply, and a very important consequence of the "locking up" of precipitation in the accumulating ice caps is a reduction in the amount of meltwater that flows back into the sea after falling as snow. As the ice caps build, sea level falls; and as sea levels go down, coastlines advance and dry land appears where before there was shallow sea. Levels may fall by as much as three hundred feet or more, with the result that, for example, during cold periods Britain was part of the European landmass, North America and Asia were joined across the Bering Strait, and Borneo belonged to Southeast Asia. Today Java is an island, possibly the last place one would expect to find an early human with African antecedents who lacked the technology to cross water; but about a million years ago *Homo erectus* was evidently able to walk there.

The earliest Europeans

Whether *H. erectus* ever made it into the then rather hostile climes of Europe is a point that has been debated endlessly. A few early archaeological sites in Europe that may date from about a million to 500,000 years ago have yielded crude stone tools. One hotly contested site at St. Eble in France has even been alleged to produce rudimentary stone tools about 2 million years old. Unfortunately, "early" European sites either have pretty firm dates but poor archaeological evidence, or good tools supported by questionable dates. And in any event, at none of them is there any direct evidence of the presumed toolmaker—which might (or might not) have been *H. erectus*.

Nobody, however, is seriously basing the argument for the presence of *H. erectus* in Europe on the rather flimsy early evidence of crude tools. Such claims are based instead mainly on the earliest well-documented human fossils that came from Europe. These are plainly not *Homo sapiens*, but since tradition admits no intermediates between the two, it seems logically to follow that they must be *H. erectus*. This is the solution generally preferred by continental European paleoanthropologists. But it's not that simple, for these

The cave site of Arago, near Tautavel in southern France. Over 400,000 years ago early humans inhabited the mouth of this cave, which overlooks the valley of the Verdouble River. Today the view is of vineyards; in colder times the valley teemed with reindeer, the principal food source of the people of Arago.

Photo by Willard Whitson.

fossils are equally, plainly not *H. erectus*. Which is why most British and American experts regard them as *H. sapiens*. To do them justice, these English-speaking experts are careful to label these fossils as "archaic *Homo sapiens*" to distinguish them from modern humans, just as continental Europeans sometimes refer to them as "advanced *Homo erectus*." But if you include a perfectly distinct species of human within another one, however careful you are to point out that it's somehow not typical, you are virtually denying it its own identity. And an identity is clearly needed here. These aboriginal inhabitants of Europe plainly belong to a separate species, descended from *H. ergaster*.

The appropriate name for this species is probably *Homo heidelbergensis*, and that's the name we will use here. This name derives from the first such fossil found, a complete lower jaw excavated in 1907 from a gravel pit at Mauer, near Heidelberg in Germany. Like many other European sites, Mauer is frustratingly difficult to date. Datable volcanic rocks

The 400,000-year-old face of a young adult *Homo heidelbergensis* from the cave site of Arago, France, with some of the tools associated with these early people at the site. The tools are relatively crude and range from large, rudimentary "chopping tools" to bifacial points and small flake tools of various kinds.

Photo by Willard Whitson.

are few and far between, and most European finds have been made either in caves or in river gravel deposits, both of which are notoriously hard to date. An educated guess based on the fauna associated with the jaw would place it at about half a million years old, which makes it the oldest reasonably dated human fossil from Europe (though an as-yet-undescribed jaw has recently been discovered in formerly Soviet Georgia which may be twice as old or more).

Unfortunately, isolated lower jaws are not the most informative kind of human fossil. They appear to vary much less among related species than other parts of the skull. But it is plausible to associate the Mauer fossil with the better-known early humans from the site of Arago, in southern France. This is a cave high on a steep valley face, whose sheltered entrance was occupied for a considerable time. Among some 50 early human fossils dated to about 400,000 years ago, a time of quite severe cold, there is a face of a young adult male and some other bits and pieces from the same individual. These showed that he had bulky brow ridges, a receding forehead, a skull capacity of about 1,160 milliliters, and largish teeth compared to later humans. Stone tools from this living site are crude, consisting mainly of flakes and choppers not too different from those of _Homo habilis_, with rather few hand axes—no technological innovation here.

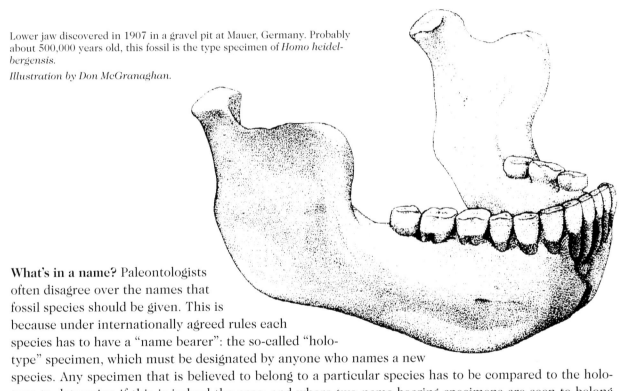

Lower jaw discovered in 1907 in a gravel pit at Mauer, Germany. Probably about 500,000 years old, this fossil is the type specimen of _Homo heidelbergensis._

Illustration by Don McGranaghan.

What's in a name? Paleontologists often disagree over the names that fossil species should be given. This is because under internationally agreed rules each species has to have a "name bearer": the so-called "holotype" specimen, which must be designated by anyone who names a new species. Any specimen that is believed to belong to a particular species has to be compared to the holotype to determine if this is indeed the case; and where two name-bearing specimens are seen to belong to the same species, the first-named has priority. Where a holotype is a good and complete example of its species, there will be few problems. But when a fossil holotype represents only a part of a skeleton, or where the biological species it represents was very variable—for example, the males and females were of very different sizes—major difficulties can result. For example, the name-bearer of _Homo heidelbergensis_ is an isolated lower jaw. We do not know what the entire skull it came from looked like. So how do we determine if the Petralona cranium, which has no lower jaw, actually belonged to the same species and should thus have the same name? A similar question arises with the holotype lower jaw of _Homo ergaster_ and crania such as KNM-ER 3733. Although the finding of associated crania and mandibles may help in some cases, the fact remains that many holotypes are unsatisfactory representatives of their species and that problems of naming are likely to multiply as more specimens are discovered and named.

A more complete skull, very poorly dated but conceivably of around the same age, is known from a cave at Petralona, in Greece, and is very comparable in known features. Its cranial capacity is about 1,200 milliliters, its face is large, and its nose is broad. Despite projecting somewhat at the rear, the braincase looks more rounded and better "inflated" than those of *H. erectus* specimens, even of comparable brain size, and this is probably why many English-speaking paieoanthropologists prefer to call such early humans "archaic *H. sapiens*."

The earliest shelters

Although it has yielded no human bones, the southern French site of Terra Amata is contemporary with these fossils and gives us an incomparable glimpse into the lives of these early people. Today perched high above the sea overlooking the bay of Nice, 400,000 years ago Terra Amata was a beach lapped by the waves of the Mediterranean. On this beach was a campsite to which at least one group of nomadic early humans returned on several occasions, creating new occupation layers on successive visits. For the first time in the archaeological record, the remains of several shelters were found, outlined in the beach sands by the beach cobbles removed to their periphery, and by the occupation litter inside. Archaeology has been accurately called the study of ancient garbage, the junk that people leave behind them, and this is certainly true of the site at Terra Amata, where the interiors of the shelters were scattered with broken animal bones, charcoal, and worked stones. The best preserved of them also contained the remains of a hearth, a ring of stones a yard across on which a fire had been maintained.

This shelter was a simple affair, although surprisingly large. It was an oval structure over twenty-five feet long and fourteen feet wide, made of saplings embedded in the ground around its periphery, reinforced with stones, and drawn together at the top. A narrow (two to three feet) gap in the ring of reinforcing stones indicates where the entrance was; at this point, too, the detritus from

Artist's reconstruction of one of the 400,000-year-old huts excavated at Terra Amata, southeastern France. Made from saplings embedded in the ground and brought together at the top, this hut probably was not waterproofed with hides. In this view the side is cut away to show the hearth (the shallow depression at lower right of cutaway) found inside, together with the debris from stoneworking.

Drawing by Diana Salles, after a reconstruction by Henry de Lumley.

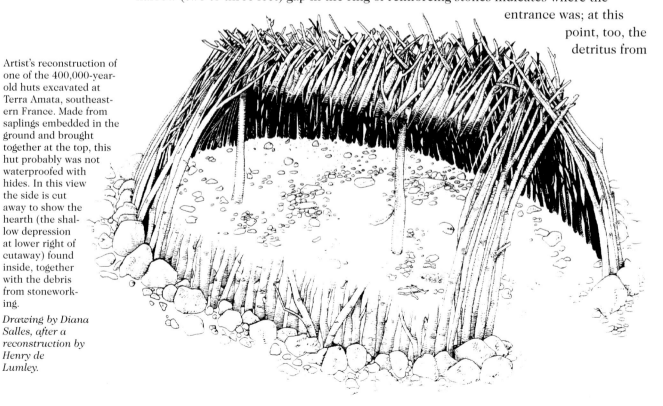

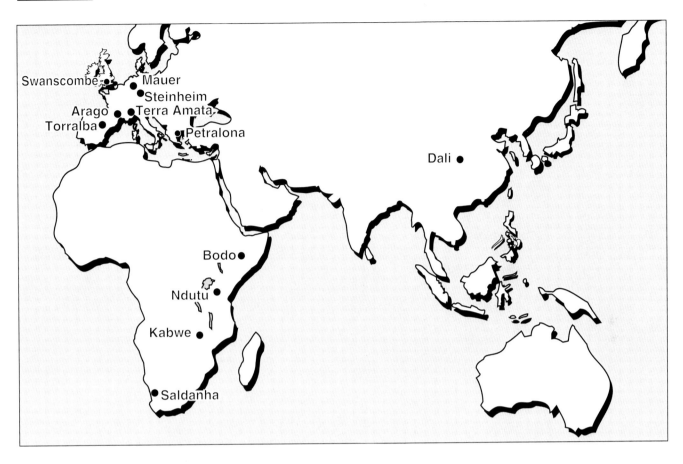

Sites where the various "intermediate" human fossils discussed in this chapter were found.

Illustration by Diana Salles.

the interior spilled over the threshold to the outside. It's impossible to know if this structure was once covered with hides to make it waterproof, though its excavator, Henry de Lumley of the Institute of Human Paleontology in Paris, thinks not: the natural dense foliage of the saplings would probably have provided adequate shelter even after it dried, and he thinks of the hut as a windbreak rather than as a perfectly waterproof shelter. But there is, of course, no way of knowing for sure.

The people of Terra Amata used local beach cobbles as the materials for a range of stone tools that, typically for its time and place, was rather crude. This may, of course, have been due simply to the fact that the local limestone is not particularly well adapted for fashioning such utensils. According to de Lumley, the huts of Terra Amata were simple short-term campsites that were inhabited seasonally by hunters who roamed through the surrounding area. Once more, though, we can't be sure quite how expert these hunters were. The animal bones they left behind included those of elephants, which roamed Europe at that time, wild boar, deer, and wild cattle. There were plenty of rabbits, too, and while we can be pretty confident that they caught such small prey, it's a little hard to imagine that they tackled the elephants directly; more likely, they came across their cadavers during their wanderings. A pair of roughly contemporaneous and adjacent sites in Spain, Torralba and Ambrona, were once hailed as places where early humans drove elephants into swamps, where the helpless pachyderms were killed; but nowadays, once again, the association of stone tools with the elephant bones appears more fortuitous than had been thought.

Advanced humans in Africa

Unfortunately, the Terra Amata site was encased by the foundations of an apartment building before any human remains had been found there. But our knowledge of the kind of early people who camped there does not stop with the few European fossils we've mentioned. Interestingly, perhaps the best-preserved individual of this kind of human comes not from Europe at all, but from the site of Kabwe (formerly Broken Hill) in Zambia, southern Africa. This fossil, known as "Rhodesian Man," was found at a site since destroyed by mining. Its date can only be guessed, though it is almost certainly in excess of 125,000 years old.

The fossil consists of a nicely preserved skull and a few bones of the body. The latter are fairly modern-looking, if robust, and suggest that in life the young adult stood about five feet, eight inches tall. The skull has a large, roundish braincase with a capacity of 1,280 milliliters, getting pretty close to the modern average (the modern range is about 1,000–2,000 milliliters, though the extremes of that range are very rare). The face was quite large, however, and was hafted somewhat in front of the braincase. It also was overhung by massive brow ridges, behind which the forehead sloped back quite sharply. Although we normally consider dental decay to be a modern affliction, Rhodesian Man had a case and a half: out of 15 teeth partly or wholly preserved in the skull, 12 are decayed, some extremely badly.

Similar human remains come from sites throughout Africa. A skullcap from Saldanha, near the continent's southwestern tip, shows similar features, and a very robust face and partial braincase from Bodo, in Ethiopia, is quite comparable despite its remarkably heavy build. The Bodo skull may be well over 200,000 years old, and the Saldanha specimen is probably a good bit older yet, at close to 350,000 years.

The very distinctive human species *H. heidelbergensis* was thus clearly a highly successful form, and it may have been even more successful and widespread than we realize if some specimens from China, such as a more or less complete cranium from the site of Dali, represent an eastern variant of this species. Unfortunately, adequate descriptions of the Chinese fossils have been extremely slow to appear, so it's hard to know for sure.

Two African crania of *Homo heidelbergensis*. On the left is "Rhodesian Man" from Kabwe, Zambia, with a brain volume of 1,280 milliliters and probably somewhat older than 125,000 years. On the right is a cranium from Bodo, Ethiopia, perhaps more than 200,000 years old. Scratches in its orbits suggest that this skull may have been defleshed after the death of its owner.

Photo by Willard Whitson.

Better stone tools

At some time before 200,000 years ago, and thus near the end of the presumed span of *H. heidelbergensis* on Earth, a new stoneworking technology appeared, first in Africa, later in Europe. Known as the prepared-core technique, it involved the careful fashioning of a lump of stone in such a way that a single final blow would detach a flake of predetermined size and shape. This flake could then be retouched, if necessary, to produce the finished tool. The great advantage of this new method was that it provided long cutting edges along the sides of the flake, plus greater control over the shape of the final tool, and it laid the groundwork for the great technological advances in toolmaking that were to come.

Also from the time of *H. heidelbergensis* comes our first direct evidence for the manufacture of wooden implements. As we've seen, indirect evidence for this activity comes from quite early African stone tools in the form of wear polish; but thanks to unusual conditions of preservation, the 300,000-year-old site of Clacton, in England, has produced an actual wooden spear point, made from yew.

A prepared-core flake tool, with the core from which it was detached. The preparation of the core ensures that the final flake tool will have a long, sharp, continuous cutting edge.
Replica tool by Dodi Ben-Ami;
Photo by Willard Whitson.

The making of a prepared-core tool. A core is first shaped to the desired form, and then a large flake is detached from it with a single blow. This flake is in essence a finished tool.

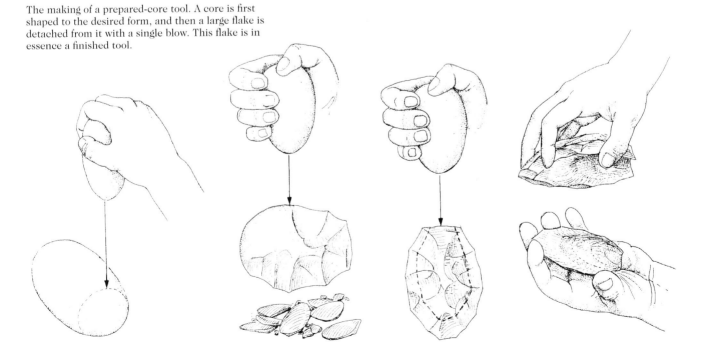

"Problem" fossils

In the period between 350,000 and 120,000 years ago we also encounter certain early human fossils that do not fit comfortably within the species *H. heidelbergensis*. Neither can they be described convincingly as *H. sapiens*, though some of them do share some features with the later Neanderthals, a distinctive group of humans known back to 200,000–150,000 years ago. Among these possible proto-Neanderthal fossils is the back of a skull from Swanscombe, England (about 200,000–250,000 years old and with an estimated brain volume of 1,325 milliliters) and a badly distorted cranium from Steinheim, Germany, which is of about the same age. Some other specimens, though, are less easily placed. A 350,000-year-old cranium from Ndutu, in northern Tanzania, is entirely distinct from the Rhodesian-Saldanha group. Clearly, during the middle Pleistocene a lot was going on in human evolution that we perceive dimly, if at all. Nonetheless, these developments are of the greatest significance, for they laid the groundwork for the emergence of our own species.

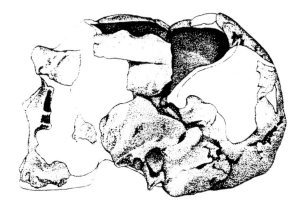
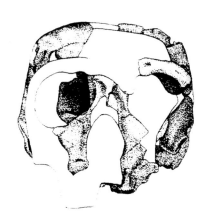

Incomplete cranium, about 350,000 years old, from the Masek Beds at Lake Ndutu, Tanzania. This specimen is unlike other "intermediate" fossils from Africa and has been compared to the somewhat younger Steinheim cranium from Germany.

Illustration by Don McGranaghan.

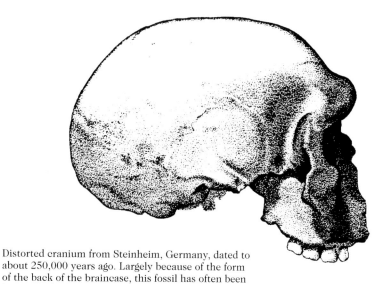
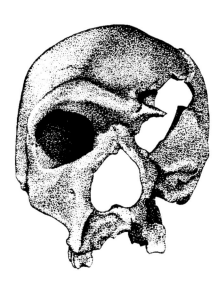

Distorted cranium from Steinheim, Germany, dated to about 250,000 years ago. Largely because of the form of the back of the braincase, this fossil has often been interpreted as a proto-Neanderthal.

Illustration by Don McGranaghan

Chapter Ten
The Neanderthals

About 150,000 or 200,000 years ago, Europe and western Asia witnessed the emergence of a distinctive new kind of human. The group is known informally as the Neanderthals, after the German site in the valley ("thal") of the Neander River that in 1857 yielded the first such fossil to be described.

The Neanderthals were more muscular and robustly built than humans today, males averaging about five feet, six inches tall but weighing about 150 pounds and up. Otherwise they differed from us in only a few details of the body skeleton, and were most emphatically not the shambling brutes depicted in cartoons. And although their skulls were very differently constructed from our own, they had brains that were on average at least as large as ours are today. Indeed, a sampling of Neanderthal skulls from Europe produces an average brain volume of about 1,500 milliliters, 100–150 milliliters above the modern average. The shape of those big brains was, however, a little different from our own, and they were enclosed in a long, low braincase that protruded at the back. Overhanging the face were substantial arching brow ridges, and the face itself was robust and forwardly placed, jutting in the midline but broad-nosed and with curiously swept-back cheekbones.

The earliest probable Neanderthal fossil is a broken braincase from Ehringsdorf, in Germany, dated to 200,000 years ago or perhaps even earlier. However, the reconstruction of this specimen is disputed, and for better evidence we have to move up to about 150,000 years ago. The French site of Biache is of this

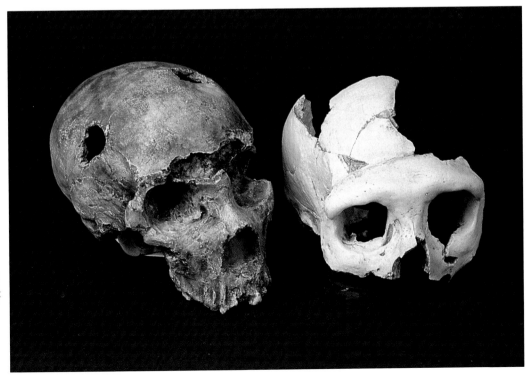

Two relatively early ("pre-classic") European Neanderthal crania from the end of the last inter-glacial, about 100,000 years ago. The specimen on the right is from the cave of Krapina, Croatia, which has yielded a large series of Neanderthal fossils spanning several millennia; the skull on the left is from Saccopastore, Italy, and has a brain capacity of 1,250 milliliters.

Photo by Willard Whitson.

age, and it has produced a partial braincase that shows the characteristic protruding occipital "bun." Neanderthals were thus around in Europe at the end of the second-to-last major cold period, but they start to become better known only during the succeeding warmer interglacial, between 125,000 and 90,000 years ago. Neanderthal remains from this time dot Europe from the Atlantic coast to Croatia, and possibly further east as well.

The "classic" Neanderthals

The heyday of the Neanderthals came during the next glacial phase, from about 90,000 to 35,000 years ago. Sites at which the Neanderthals of this period camped are known widely from France and Spain all the way east to Uzbekistan. Western European Neanderthals of this period show the same basic physical features as those found earlier and further east, but in a rather more pronounced form. Because of their distinctive anatomies they are known as "classic" Neanderthals, and it is popular to explain their appearance in terms of adaptation to the extreme glacial conditions in which they lived. The broad nose, for instance, is widely thought to be an adaptation for warming and moisturizing bitterly cold, dry air before it entered the fragile lungs. The archetypal classic Neanderthal is the skeleton of the "Old Man of La Chapelle," discovered at a site in France in 1908. It was the analysis of this skeleton by the distinguished French anatomist Marcellin Boule (1861–1942) that did much to establish the idea of Neanderthals as bent-kneed, shuffling creatures: only much later was it realized that many of the features of the skeleton to which Boule drew attention were due to the fact that the aged and unfortunate individual had suffered from an advanced case of arthritis.

Two French "classic" Neanderthal crania, from the last glacial period. The fossil on the right is the "Old Man" of La Chapelle-aux-Saints, early studies of whose aged, arthritic skeleton gave rise to the myth of the shambling, bent-kneed Neanderthal. It is about 50,000 years old, with a large brain volume of 1,620 milliliters. On the left is a cranium from La Ferrassie, perhaps a little younger and with an even larger cranial capacity of 1,680 milliliters.

Photo by Willard Whitson.

Sites that have yielded Neanderthal fossils. Early Neanderthal sites from the last interglacial are shown by open circles, later ones by solid dots. *Illustration by Diana Salles.*

Who were the Neanderthals?

When the first Neanderthal was described in 1857—two years before the publication of Darwin's *On the Origin of Species*—the interpretation of this big-brained but distinctive human type posed considerable difficulties for the scientists of the day. For in those pre-evolutionary times the notion that other human species had existed before modern humans came on the scene was far from people's minds.

It's not surprising, then, that most scientists tried to explain this apparent anomaly as some odd kind of modern human. One school of thought held that the Neanderthal specimen represented a vanished "barbarous race." With Gallic disdain, for example, the French anatomist Franz Pruner-Bey (1808–1882) diagnosed the Neanderthal skullcap as that of a strongly built Celt "somewhat resembling the skull of a modern Irishman with low mental organization."

More popular was the idea that the Neanderthaler was the remains of some kind of pathological modern human. The German anatomist Friedrich Mayer (1787–1865) took this notion to its extreme when he ingeniously put together three false observations. The brow ridges, he claimed, resulted from a constant frown caused since childhood by the pain of rickets; the thigh bone showed a curvature typical of horsemen; and the braincase showed some Mongolian features. In combination, these three notions suggested to Mayer that the skeleton had belonged to a sick Cossack who had deserted from the Russian invading force of 1814! This sort of thing didn't go down well with everyone: not long afterward the examination of a similar skull found in Gibraltar in mid-century allowed George Busk (1807–1886) of London's Royal College of Surgeons to write, with evident satisfaction, that "Even Professor Mayer will hardly suppose that a ricketty Cossack engaged in the campaign of 1814 had crept into a sealed fissure in the Rock of Gibraltar!" Nonetheless, it wasn't until well into the twentieth century before accumulating Neanderthal finds made it obvious to everyone that here was something that was not just an odd type of modern human.

Of course, we can hardly blame the anatomists of the mid-nineteenth century for interpreting the bizarre Neanderthal find in terms of the world they knew. But even today the fanciful explanation of sparse facts about the Neanderthals is reinforced by their traditional interpretation as an odd variant of our own species *Homo sapiens*. For if they were indeed *H. sapiens*, they were to all intents and purposes fully human and could be expected to indulge in all of the bizarre and complex behaviors typical of our kind. Holdover from the nineteenth century though it is, it's still probably the majority opinion among paleoanthropologists that the Neanderthals were actually a variant of our own species, the subspecies *Homo sapiens neanderthalensis*. More and more of them, however, are arriving at the conclusion that the Neanderthals did indeed belong to a species distinct from our own, and should thus be known as *Homo neanderthalensis*. This species was probably descended, like ours, from *Homo heidelbergensis*. But it was not our ancestor.

Burials and "bear cults"

The proposed evolutionary relationships of later human species. Shaded bars depict the documented time ranges of the various species, and the lines represent potential relationships among them.

Illustration by Diana Salles.

Most Neanderthal remains have been discovered in or near the entrances of caves, not simply because these were favorite spots for sheltering, but also because these people buried their dead in such places—apparently the first human beings to do so. In a provocative attack on received wisdom the archaeologist Robert Gargett has, however, recently cast doubt on the whole notion of Neanderthal burial. "Processes other than purposeful human behavior," says Gargett, "may have produced the deposits in question." Nonetheless, while Gargett is undoubtedly correct in claiming that the elaborateness of many or most such burials has been highly exaggerated, there can be little doubt that at least occasionally, and very simply, Neanderthals did bury their dead companions.

Exactly what this innovation implies about the Neanderthals' sensibilities is not clear, though everyone can agree that such behavior testifies to some form of self-aware-

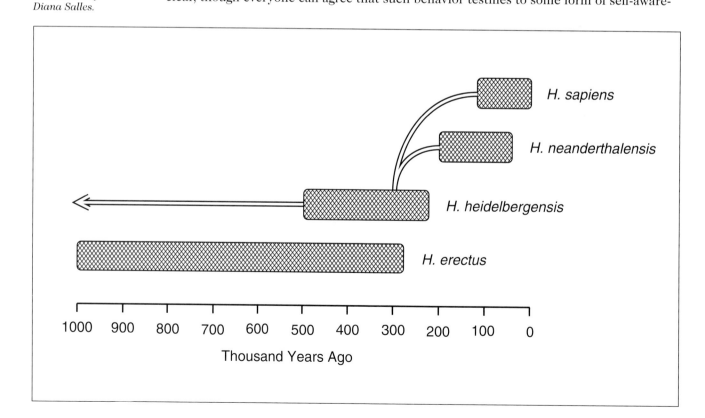

ness and spiritual feeling. But lurid stories of obscure religious rites and of Neanderthal "bear cults"—the worship or sacrifice of the enormous cave bears common in Ice Age Europe before about 50,000 years ago—are evidently untrue, or at least are not sustained by reliable evidence. Such stories stem mainly from early excavations undertaken when archaeology had not yet developed into the meticulous science it is today, and when the reconstruction of early ways of life had as much to do with the imaginations of the excavators as with the facts at their disposal—many of which were obscured anyway by deficient excavating techniques.

For example, recent reappraisals have shown that the apparent careful placement of a Neanderthal skull within a ring of bones at the site of Monte Circeo in Italy is not borne out by careful analysis of the evidence available. Neither is the story of the bear cult of the Drachenloch cave in Switzerland, which is based on a fanciful interpretation of facts that are easily explained by natural causes. Both interpretations turn out to have been based on reconstructions made many years after the fact by investigators who may not even have been present when the bones were actually discovered by the laborers who cleared the sites for them.

Neanderthal tools and lifestyles

What does the archaeological record tell us about the behavior of the Neanderthals? What do we actually know about how they lived their lives? Perhaps most obviously, they were magnificent toolmakers. Their "Mousterian" tools, named for the site of Le Moustier in southwestern France, represented a refinement of the basic prepared-core technique. More than sixty different types of Mousterian flake tools have been named, together with some twenty different kinds of small hand axes made from cores. All of these are carefully shaped with a high degree of precision; but beautiful though these tools are, as the eminent French prehistorian François Bordes once remarked, they were made "stupidly." By this he meant that Mousterian tools lacked the spark of individuality and ingenuity that characterized the products of early modern humans in Europe. The Neanderthals produced Mousterian tools over a vast area of Europe and western Asia; but wherever these implements come from, they look pretty much the same. Careful copying, rather than individual creativity and imagination, seems to have been the rule among the Neanderthals, especially since there is no substantial change in the tool kit through the long Mousterian period. It's perhaps significant as well that Neanderthals seem never to have imported stone to make their tools from more than a few miles away from where the implements were found: in this, too, they contrast greatly with the humans who succeeded them.

Neanderthal sites are mostly found in cave entrances and beneath rock overhangs. Such places provided airy natural shelters for these people, and also environments conducive to preservation of their leavings. But Neanderthals camped in the open also, and even rigged up skin-and-pole shelters within the cave mouths they inhabited. Most of the known sites were not one-time camps but semipermanently inhabited living places, visited intermittently over years. It's hence a bit surprising that cooking hearths, well known in the Middle East, are not common from Neanderthal sites in Europe, although burned bones and charcoal provide plenty of evidence for the use of fire.

The wear on Mousterian stone tools indicates that some were used to shape wood. Indeed, the sharpened point of a ten-foot-long spear was remarkably preserved at the German site of Lehringen, in among the fossilized ribs of an elephant. Other tools were used to scrape hides. Some of those hides must have been used for a rudimentary form of clothing, which was presumably essential for survival in the harsh conditions of the last

Life-size reconstruction of a male "classic" Neanderthal prepared for display at the American Museum of Natural History. The individual is shown sharpening a wooden spear, an activity documented from the wear on stone cutting tools as well as from rare preserved wooden spear points up to 300,000 years old. Features such as the quantity and distribution of hair, the form of the nose, and the color of the skin, are conjectural.

Photo by Dennis Finnin and Craig Chesek.

Mousterian flint tools from various sites in western France. From left: scraper on flake; two small handaxes; scraper and point on flakes.

Photo by Ian Tattersall.

Life-size reconstruction of a young female "Classic" Neanderthal scraping a hide, prepared for display at the American Museum of Natural History. Studies of the wear on Neanderthal stone tools demonstrate that they were used for tasks such as this, and the unusually heavy wear typically found on the front teeth of Neanderthals suggests that they were regularly employed in the preparation of hides.

Photo by Dennis Finnin and Craig Chesek.

glacial period. In addition, the front teeth of most Neanderthals are very heavily worn, and among modern peoples such extreme wear is usually caused by chewing hides to make them softer. Most of the animal bones found at Mousterian sites are those of medium-to large-sized mammals such as deer, bison, reindeer, wild cattle, and horses; the bones of woolly rhinos and elephants, the largest mammals of the time, tend to be lacking. Archaeologists once again decline to be very specific about what these bones imply about the Neanderthals' hunting prowess, though it's clear that cooperative hunting was part of their behavior pattern. There's no doubt, however, that meat was supplemented by gathered plant foods, and indeed these probably constituted the majority of the diet in warmer periods when the vegetation was more varied.

An aspect of Neanderthal life that has received a lot of attention is the amount of support that was at least occasionally lavished upon individuals within the group. The classic example of this comes from the Iraqi cave site of Shanidar, which has yielded six reasonably complete Neanderthal skeletons. One of these belonged to a male whose right arm had been withered, quite possibly since birth. Nonetheless, for a Neanderthal he had lived to an advanced age, possibly in the forties. It's unlikely that this man would have survived so long without the consistent support of his group. That he did so—and maybe also that other members of the Shanidar group also had survived quite serious wounds—indicates that compassion existed in the society that sustained him. "Individuals," says Erik Trinkaus of the University of New Mexico, "were taken care of long after their economic usefulness to the social group had ceased."

There's also other possible evidence of "modern" behavior from Shanidar. A couple of crania from the site may have been artificially deformed by binding the head when the individuals were young, a practice otherwise unknown except among modern people. A

Two Neanderthal crania from the Middle East. On the right, the Shanidar 1 cranium, belonging to an aged individual whose withered arm suggests that he was supported by his group for a long period. On the left is a cranium from the cave of Amud in Israel. About 50,000 years old, both of these fossils were contemporary with the "classic" Neanderthals of western Europe, though they were not quite so highly specialized in cranial features; both had large brains (about 1,600 milliliters and 1,740 milliliters, respectively).

Photo by Willard Whitson.

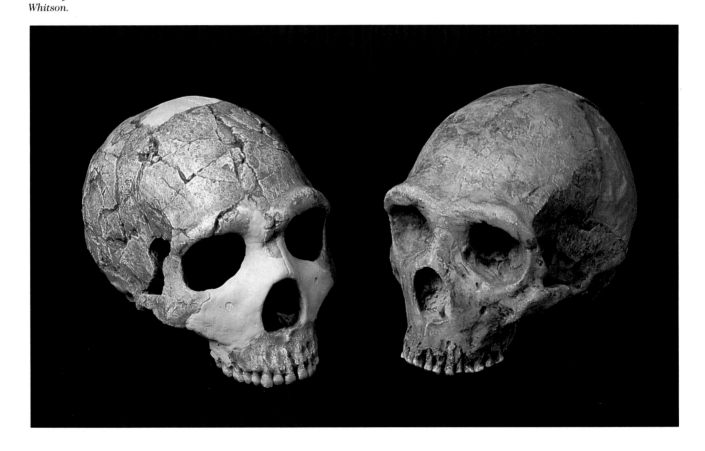

Views of the Upper Shelter site at Le Moustier, south-western France. This site was occupied by Neanderthals over a considerable period, and gave its name ("Mousterian") to the stone tool kit typically associated with Neanderthals in this part of the world. Overlooking the Vézère River, this site was a superb lookout for hunters preying on the herds of reindeer and other mammals that grazed on the floodplain below.

Above: looking out from the shelter, across a terrace above the modern hamlet of Le Moustier.

Below: view toward the shelter from a rock overhang partway up a limestone bluff.

Photos by Willard Whitson.

further humanlike Neanderthal practice that has been pointed out is the carrying home of ochre and similar pigments. Since there is no evidence that these colors were used to decorate objects, it has been suggested that they were employed for body painting. It's just as likely, however, that they were used for more prosaic purposes, such as treating hides or wounds. What's more, Neanderthals never heat-treated such pigments, as early modern Europeans did.

How "human" were the Neanderthals?

We have, then, some small suggestion of "modern" behavior patterns among the Neanderthals. But looking over all the various lines of evidence, it's hard to conclude that they were doing most of the things that we would intuitively recognize as "human." They might, indeed, best be described as having done what their precursors had done, only a little better. The same could be said of all of the previous originators of the cultural changes mirrored in the archaeological record. No question, in its 5 million years or so the human lineage had come a long way; but right through the time of the Neanderthals, innovations had been isolated and episodic, and none of them had really amounted to a total revolution in daily life. This pattern of occasional small changes contrasts significantly with the archaeological record left behind by the Neanderthals' successors in Europe. In this comparison the Neanderthals suffer, maybe unfairly.

Chapter Eleven

The Origin of Modern Humans

The Neanderthals vanished abruptly from western Europe shortly after modern people arrived. These modern people were appreciably different from the Neanderthals from the neck up. They had higher, shorter, and more lightly built braincases that were more or less rounded at the back. In front a more or less vertical forehead rose above small or nonexistent brow ridges. Their faces were small and were tucked below the front of the braincase. They had chins, in place of receding jaw profiles. They were on the tall side, with lightly built bodies. They were *Homo sapiens*.

The appearance on Earth of modern people was a major event in human evolution, yet it remains among the most poorly understood of such events. This is partly because the behavior of *H. sapiens* did not immediately differ dramatically from that of its predecessors, although it certainly came to do so. And it is also partly because the fossil evidence for the appearance of modern humans is sparse and difficult to interpret. Most important of all, however, is the influence that notions about how evolution works have had on the reading of the human fossil record. And paleoanthropologists are deeply divided about how evolution occurs. One school of paleoanthropology, led by

Diagram contrasting the "regional continuity" model of modern human origins with the competing "center of origin" model. In the former case the various regional variants of *Homo sapiens* are seen as evolving in parallel from *Homo erectus* while remaining united by liberal gene exchange among them. The "center of origin" model conforms better to what we know about how new species originate; in the interpretation offered here, *Homo heidelbergensis* gave rise on the one hand to the now extinct species *Homo neanderthalensis*, and on the other to our own species, *Homo sapiens*, which diversified subsequent to its origin (probably) in Africa.

Illustration by Diana Salles.

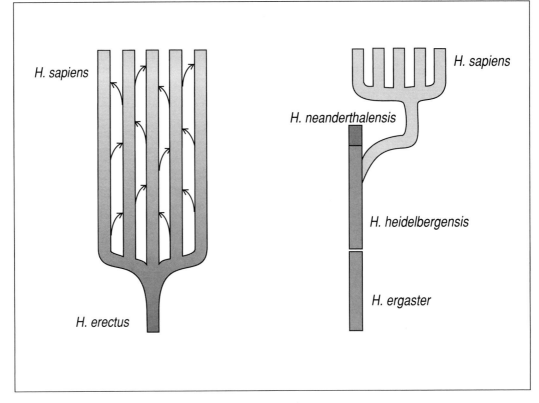

133

Milford Wolpoff of the University of Michigan, adheres strictly to the idea of gradual change in evolution. These scholars look in the fossil record for evidence of the tiniest gradations from one kind of fossil human to another over time. Another group of paleoanthropologists, of whom I am one, is mainly concerned with the origins of new species, and thus is more sensitive to discontinuities in the fossil record.

The first kind of outlook has given rise to the so-called "multiregional" model of *H. sapiens* evolution. Wolpoff and his colleagues suggest that over the past half-million years or so, local populations of humans were in enough contact with other groups that they avoided splitting into different species. At the same time, however, these populations remained sufficiently isolated to accumulate adaptations to local environments. In this way the varieties of modern humans would have evolved their differences over long periods of time, while staying part of one large interbreeding population: one species. Thus, for example, local Chinese *Homo erectus* is seen as giving rise to today's Chinese through a continuous process of change, while the native inhabitants of Australia were derived from later forms of Java Man. Throughout the Old World, such evolving lineages remained in sufficient reproductive contact with each other to remain members of one single species.

Other scientists look for evidence of the emergence of *H. sapiens* not worldwide but in a particular place: a center of origin. This is because new primate species arise when a single species population is split into two by some kind of ecological or geological event that interrupts the free flow of genes. The newly isolated populations are then free to develop the genetic mechanisms that will ultimately prevent successful interbreeding between them should they be reunited. Such isolated populations must inevitably be associated with a particular tract of territory, and it is this that

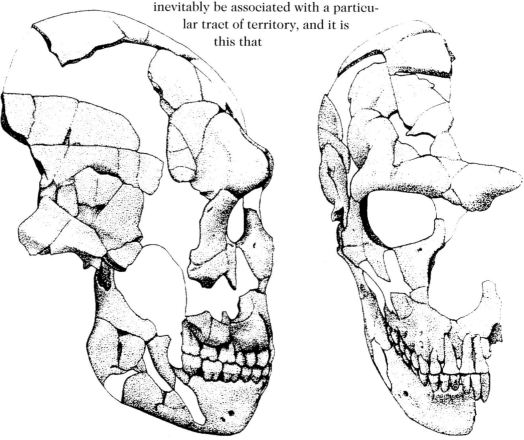

The partial skull of a Neanderthal found in 1979 at St. Césaire, France. Approximately 36,000 years old, it is about the latest definitely dated Neanderthal. The St. Césaire individual was associated with Châtelperronian implements and seems to prove that it was indeed Neanderthals who were responsible for this "intermediate" culture.

Illustration by Don McGranaghan.

will correspond to the center of origin of the new species. If it is successful, the species will spread out from that center when changing ecological or geographic conditions permit, just as our species *H. sapiens* did in eventually colonizing virtually all of the inhabitable regions of the world.

My view is that both the fossil evidence and the genetic structure of modern populations fit much better with the center of origin idea for modern humans than with the multiregional one. What's more, it seems pretty clear that once again, as for so many other innovations in human evolution, we have to turn to the continent of Africa in seeking that center of origin. However we view the evidence, though, we must examine it region by region. Let's start with Europe, because even if it's plain that this was not the site of the emergence of modern humanity, the fossil record there is both good and quite straightforward to interpret.

Europe: Neanderthal Armageddon?

Neanderthals flourished in western Europe until about 35,000 years ago, when they were replaced over at most a few thousand years by humans of modern type: *H. sapiens*. We see this replacement in the living sites that people of both kinds left behind. Archaeological sites that were occupied for any length of time consist of layers built up one on the other, as more-recent people deposited their leavings on top of those of earlier times. Sites exist in France where layers of Neanderthal occupation have been found lying above and in-between layers that consist of the debris left by modern humans. This shows that the replacement of the Neanderthals in this part of the world was not an instantaneous event, but it's nonetheless evident that the time of coexistence or alternation between these two kinds of human was not a very long one.

One difficulty in understanding this transition has been the general lack of fossils in the critical period of replacement. Much of the evidence by which we understand this episode thus comes from stone and other tools. Interpretation of these is, however, complicated by the appearance in some places of a technology that is somewhat intermediate between the Mousterian or Middle Paleolithic tools of the Neanderthals, and the Upper Paleolithic tools of the new people. Let's look at this a little more closely.

There is a major difference between the stone tools of the Middle and Upper Paleolithic. As we've seen, the Neanderthals

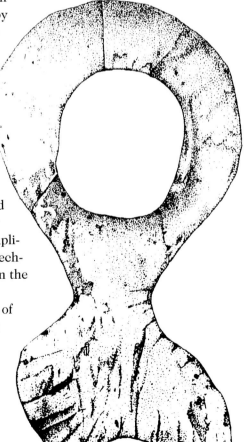

A carved bone pendant from Arcy-sur-Cure, France. This is one of the very few decorative items to have been found in a Châtelperronian context, and hence to be apparently associated with Neanderthals.

Illustration by Diana Salles.

used a refinement of the prepared-core technique, whereby a flake of specific size and shape was struck off a preformed core. Minor shaping, or "retouch," then transformed this flake into a finished tool. The first modern Europeans, in contrast, brought with them blade-based technologies. Here a single cylindrical core was prepared in such a way that a series of long, narrow flakes, called blades, could be detached from it by a succession of blows from a hammerstone, or, more likely, from a softer implement made of bone or antler. These blank blades were then modified by retouch into a wide variety of tools. The Upper Paleolithic was also characterized by the making of tools in a variety of previously largely ignored materials such as bone, ivory, and antler, and by the production of decorative objects.

In France and northeastern Spain, between about 36,000 and 32,000 years ago, we find a tool-making industry, called the Châtelperronian, which has some of the aspects of both the Middle and Upper Paleolithic. About half the stone tools of the Châtelperronian tool kit were made from prepared cores, while others were made

Stages in the making of Upper Paleolithic blade tools. A long cylindrical core is prepared, from which long, thin blades are successively detached by single blows. These blank blades are then finally shaped by retouch according to their intended purposes.

Illustration by Diana Salles.

from blades. Further, while no decorated objects and only the rarest bone or antler tools come from any Mousterian sites, occasional polished bone pendants and beads are known from Châtelperronian localities, and bone and antler tools also were fashioned.

At only one site, however, is there any substantial association of Châtelperronian artifacts with a human who might have made them. This is at St. Césaire, in France, and the 36,000-year-old individual is definitely a Neanderthal. The discovery of this fossil in 1979 seems to have settled the longstanding argument over who made the Châtelperronian tools in favor of the Neanderthals; and the intriguing suggestion has recently been made that the Châtelperronian represents the kind of industry that Neanderthals produced once they had seen the Upper Paleolithic people strut their stuff. After all, the Neanderthals' expertise in producing fine but unvarying prepared-core tools is pretty good evidence of these peoples' individual gift for imitation!

The earliest dates for the appearance of modern people in Europe come from the East, where Upper Paleolithic sites such as Bacho Kiro in Bulgaria are dated to well over 40,000 years ago. Reliable dates for Upper Paleolithic occupation in western Europe, on the other hand, are a good deal more recent than this, falling after about 35,000 years ago. _Homo sapiens_, it thus appears, entered Europe from the East and spread westward, taking over from the Neanderthals by 33,000 or 32,000 years ago (though Neanderthals may have lingered on in the Iberian peninsula a couple of thousand years longer). The early modern Europeans were of fully modern type, if maybe a little more robustly built than the average person today.

They are totally different from the Neanderthals and had clearly evolved into _H. sapiens_ elsewhere before arriving in Europe. They did not evolve from Neanderthals. Since this didn't happen, the Neanderthals and modern people must have confronted each other in some way for at least a limited period.

What shape this confrontation took has been the subject of extensive speculation. It has been suggested that the Neanderthals interbred widely with the invaders, their genes and appearance eventually becoming swamped by those of the latter. Newspapers and magazines regularly carry stories with the theme of "do we have Neanderthal genes?" Since geneticists tell us that we share more than 99 percent of our genes with chimpanzees, the answer must in one sense be yes. Almost all of our genes must have been shared by Neanderthals. But the answer must be very different if we ask ourselves, "Are we descended from Neanderthals?" The fossil evidence usually brought out in support of a Neanderthal ancestry for Europeans and Western Asians is pretty thin at best; and if Neanderthals and modern humans indeed belonged to different species they could not have interbred effectively (although individuals might conceivably have done so, willingly or otherwise).

But if Neanderthals and modern Europeans did not interbreed on any significant scale, did they at least coexist relatively peacefully, the Neanderthals simply becoming outcompeted for resources by the smarter (for certainly not stronger) moderns over a few centuries or millennia? This seems unlikely in view of the nasty ways in which invading _H. sapiens_ populations have treated each other throughout recorded history. And although studies of modern hunting-and-gathering peoples have suggested that violence is usually low in such societies, this clearly isn't necessarily true. Research on skeletons found in ancient cemeteries of such peoples have revealed that half or more of the occupants may have died violent deaths.

Upper Paleolithic blade tools in flint, from various periods and sites in western France. From left: two points; small laurel-leaf; endscraper with point at other end; unretouched blade; awl.

Photo by Ian Tattersall.

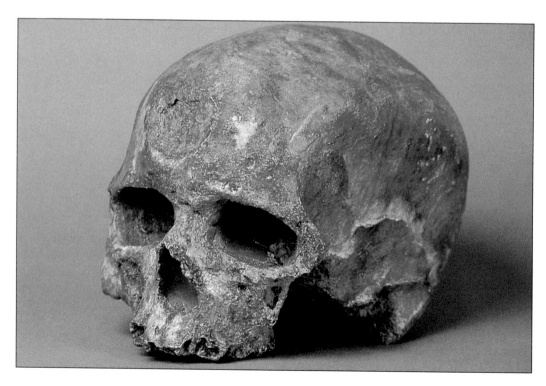

The "Old Man" skull from the rock shelter of Cro-Magnon, Les Eyzies, western France, dated to about 25,000 years ago. Discovered in 1868, the group of burials at Cro-Magnon provided the earliest uncontested evidence for the existence of modern people at the same time as a variety of extinct mammal species. Cro-Magnon ultimately gave its name to the earliest modern human inhabitants of Europe as a whole.

Photo by Willard Whitson.

The Mediterranean Basin: "ancient modern" people

New dating techniques have revealed astonishingly early dates for anatomically modern people in the part of the world that borders on the eastern and southern Mediterranean. In Israel the sites of Jebel Qafzeh and Skhūl, both initially discovered and excavated before World War II, have recently yielded dates of close to 100,000 years for human fossils that are entirely or substantially modern in appearance; and in Mediterranean Africa the site of Jebel Irhoud in Morocco, first excavated in 1962, has produced a couple of crania that are only a little more archaic-looking and which may date even earlier than this.

These dates contrast with much more recent ages for various Neanderthal sites in Israel of about 45,000 years (Kebara) to 60,000 years (Amud). A very recently obtained date, made over sixty years after excavation began, has placed a Neanderthal from another Israeli site, Tabūn, at more than 100,000 years. Such dates as these indicate clearly that in the Middle East there was a very long period of

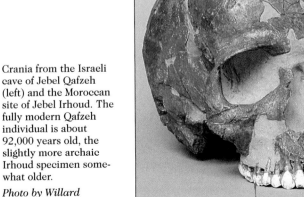

The cave of Jebel Qafzeh, just outside Nazareth, Israel. Fully modern human remains more than 90,000 years old are associated here with Mousterian industries functionally indistinguishable from those of the Neanderthals.

Photo by Willard Whitson.

Crania from the Israeli cave of Jebel Qafzeh (left) and the Moroccan site of Jebel Irhoud. The fully modern Qafzeh individual is about 92,000 years old, the slightly more archaic Irhoud specimen somewhat older.

Photo by Willard Whitson.

The circum-Mediterranean region, showing various sites (indicated by circles) that have yielded Neanderthal, early modern or near-modern human fossils.

Illustration by Diana Salles.

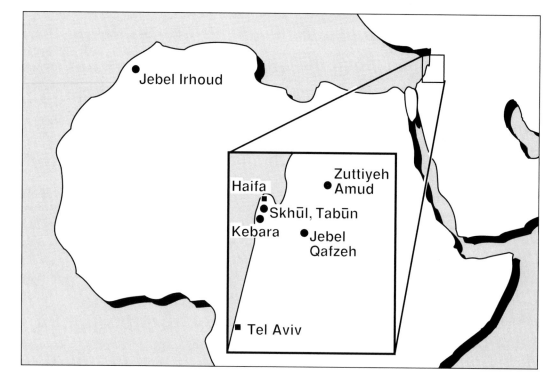

coexistence, or maybe of alternating existence, between the Neanderthals and people of modern anatomy. Evidently the short-term replacement of *Homo neanderthalensis* by *Homo sapiens* that is so clear-cut in Europe did not occur in the Near East. It may be significant, however, that the latest date for a Neanderthal in the Near East coincides pretty closely with the earliest date for the Upper Paleolithic (as opposed to anatomically modern humans) in the region.

What might account for the difference? If we follow the argument that Neanderthals were adapted to cold conditions, we might guess that they moved into the circum-Mediterranean region during cooler times, withdrawing when warmer conditions returned. The modern types, in turn, would have done the reverse. But this is no more than a guess, and the argument is weakened by the timing of the spread of *H. sapiens* into Europe. For the rapid and successful invasion of western Europe by modern people between 35,000 and 30,000 years ago took place at a time when the last glacial period was approaching its most severe point; and this hardly suggests that they were at a disadvantage in colder conditions.

Maybe, however, a cultural element enters here. One of the great puzzles posed by the Israeli sites is that the stoneworking cultures of the early moderns are effectively indistinguishable from those of the Neanderthals. These industries were not blade-based. What's more, they were bereft of the bone and ivory implements made by the earliest modern inhabitants of Europe, and there is no evidence of any decorative or symbolic activities. As we've seen, the general pattern in human evolution is for technological innovations not to coincide with the appearance of new kinds of human, so from this perspective it's not particularly surprising that the first modern people of the Levant weren't behaving very differently from the Neanderthals. But a peculiarity of humans is a desire to explain everything; and it's certainly hard to explain why the expression of what looks to us like an innate human capacity simply did not occur over a period of maybe 70,000 years or more after our species appeared on Earth.

Since differences between species are often subtle, and are sometimes confined to changes that are not obvious in the bones and teeth, one might hazard that the early anatomically modern humans simply didn't belong to the same species as we do, and that the origin of our own behaviorally overendowed species lay in a speciation event whose consequences are not visible in the skeleton. One might; but the full flowering of the human capacity first became evident not much more than 30,000 years ago, and then in Europe, a dead end in the far western corner of the Eurasian continent. If we were to take that as the starting point of our species, we would have to explain how the human capacity spread from this dead end throughout the world, where all modern human populations show the same capacity for spiritual beliefs, for self-expression, for creativity and for all the hallmarks that make us feel so different from the rest of the living world. And not only is there absolutely no reason to believe that any such spread took place, but there are plenty of reasons to believe that it didn't. No, this capacity must go back a longer way, to the ancestry of all living human populations. For which, most probably, we must look—yet again—to Africa.

Out of Africa, always something new!

There are many reasons for us to focus on Africa in the search for the origin of our own species _H. sapiens_. The earliest fossils that hint at the emergence of modern humans come from Africa south of the Sahara. Blade-based tools were being made in Africa before we know that they were being made anywhere else. And exciting new methods of genetic analysis also suggest that the earliest modern human populations may have originated there.

The genetic evidence is furnished mainly by mitochondrial DNA, or mtDNA. Earlier we saw that mitochondria are tiny structures in the cytoplasm of the cell which contain a small quantity of DNA. Since the mother's ovum is an entire cell, cytoplasm and all, while

---

DNA and evolutionary relationships. In recent years analyses of DNA have been widely used to determine the closeness of relationships between species belonging to various groups of organisms. One approach capitalizes on the double-stranded nature of DNA. If single strands of DNA from different species are "hybridized" in the laboratory, the strength with which they bind together is proportional to their similarity. This method is particularly well-known for its use in demonstrating the very close genetic relationships between humans and the African apes. Another avenue is offered by mitochondrial DNA (mtDNA: DNA from outside the cell nucleus, inherited only from the mother). Besides its unusual mode of inheritance, mtDNA is small in quantity and thus it is practical to "read" an individual's entire mtDNA, or substantial portions of it. This method has become particularly well-known for the support it has lent to the "African Eve" hypothesis that all modern humans are descended from a single female from Africa. Finally, base sequences of various specific portions of the nuclear DNA genome can be read. In such cases, specific genes may be identified and characterized. However, such studies yield base sequences that are very long, and which must be analyzed by computer making certain specific assumptions. Thus there is not only limited sampling of the genome, but also much discussion over how the data obtained are best interpreted. Nonetheless, as techniques are improved and procedural difficulties are overcome, we may expect to see molecular techniques become increasingly important in the assessment of relationships between species, and among populations within species.

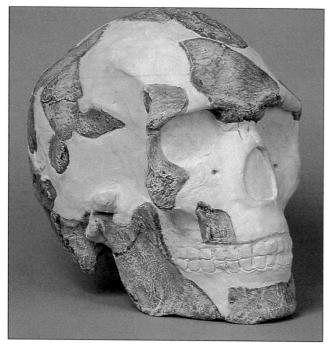

Two partial crania from the Kibish Formation, Omo Basin, Ethiopia. Both possibly about 125,000 years old. Omo I, reconstructed on the left, is perfectly modern in appearance; Omo II, on the right, is less modern looking.

Photos by Willard Whitson.

the father's sperm contains only nuclear DNA, the mitochondria that are ubiquitous in the cells of our bodies are all inherited from our mothers. Studies of human mtDNA were pioneered in the laboratory of the late Allan Wilson of the University of California, Berkeley.

By studying a sample of humans of African, Asian, European and Australasian descent, Wilson and his colleagues found that we are a close-knit species indeed: although mtDNA evolves rapidly compared to nuclear DNA, it varied rather little within and among these populations. But the people of African descent showed the most variation among themselves, and were most distinct from other populations. Since, other

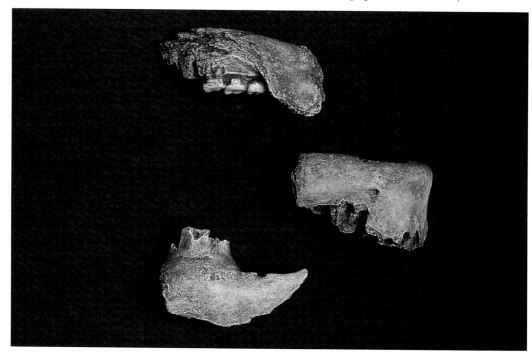

Some of the modern human skull fragments from the South African cave site at Klasies River Mouth, all probably over 100,000 years old. The charred and fragmentary condition of these fossils may be due to cannibalism.

Photo by Willard Whitson.

The cave site of Klasies River Mouth, near the southern tip of Africa. The archaeological layers are found in a number of cavities below the large hollow toward the bottom of the cliff face.

Photo by Ian Tattersall.

things being equal, a greater variety indicates a longer evolutionary path, this is precisely what you would expect if all modern humans were descended from a single population that arose in Africa. Further, making certain assumptions about the rate at which DNA changes allowed the researchers to calculate that this origin occurred about 200,000 years ago. The conclusion reached by Wilson and colleagues has predictably become quite controversial, and it has recently been attacked by such eminent geneticists as Washington University's Alan Templeton. Clearly, the last word has yet to be written on the subject. But there are other reasons for viewing Africa as the cradle of mankind, and most of them lie in the fossil record. Let's look briefly at that record.

More African fossils

In 1967 a Kenyan team led by Richard Leakey discovered two partial skulls in Ethiopia's Omo Basin. Both may be as much as 125,000 years old, although the dating of neither is firm. They make an interesting contrast, for while one is rather archaic, with a low skull with an angled rear end, the other is very modern-looking. They obviously didn't come from the same population, but without better dating it's hard to say what either of them means for the evolution of our species. A more definite date of about 120,000 years is available for a fossil braincase and upper jaw discovered in 1976 by a team led by Mary Leakey in the Ngaloba Beds at Laetoli in Tanzania. This specimen does not, however, look very modern; possibly it represents an African equivalent of the European Neanderthals, although it has less-pronounced features and is thus more plausibly ancestral to later humans. It may prove to be quite similar to a fossil face from Florisbad in South Africa that turned up when a local entrepreneur was digging out warm-water springs on his farm in hopes of creating a spa resort. As one might expect, this specimen is not reliably dated, but it, too, may be about 120,000 years old. If so, perhaps we are looking here at late-surviving forms of the ancestor of all of us. Perhaps.

Some specimens that are clearly modern in skull shape come from further east on the continent, at Border Cave on South Africa's frontier with Swaziland. It is possible that

The location of various sites in sub-Saharan Africa that have yielded fossils of early-modern or near-modern humans.

Illustration by Diana Salles.

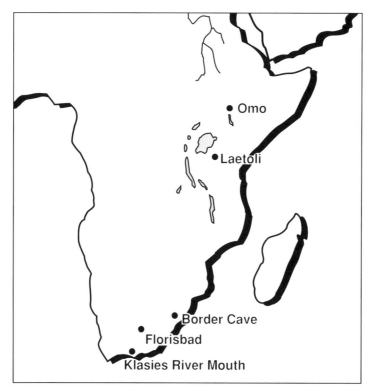

these anatomically modern humans are more than 100,000 years old, but unfortunately the best specimens were not recovered under well-controlled conditions, and the level from which they came rests in doubt. Close to the southern tip of the continent, however, we encounter fossils that clearly are those of anatomically modern people, and that almost certainly do date to more than 100,000 years ago. Indeed, some may be considerably earlier, dating back perhaps to 120,000 years.

These specimens come from caves at the mouth of the Klasies River first excavated by Ronald Singer and John Wymer between 1966 and 1968. They consist of a number of fragmentary bits and pieces that, despite their incompleteness, are without any doubt whatsoever the remains of anatomically modern humans. Once again the archaeological context is Middle Stone Age, the African equivalent of the Mousterian, but Hilary Deacon of the University of Stellenbosch, the site's most recent excavator, finds hints at Klasies of modern behavior. "Archaeological evidence," says Deacon, "shows that people in that time ranged around organized domestic hearths, using resources of both the land and sea, having rules of

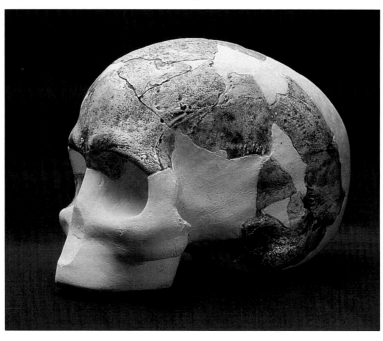

Reconstructed skull of a modern individual from Border Cave, on the frontier between South Africa and Swaziland. This individual may be as old as 90,000 years, although this date has been questioned.

Photo by Willard Whitson.

cleanliness and investing artifacts with symbolic meaning." More ominously, perhaps, Deacon sees evidence to suggest that the fragmentary and burned condition of the human fossils is due to cannibalistic activities. If so, this is the earliest well-documented evidence of such behavior among modern humans.

At Klasies River Mouth, then, we have the best suggestion

for the presence of modern humans maybe as much as 120,000 years ago—a demonstrated antiquity as yet unmatched elsewhere in the world. As time passes, the huge continent of Africa will, with luck, provide for paleontologists more of the surprises which almost 2,000 years ago led Pliny the Elder to exclaim: *"Ex Africa semper aliquid novi!"* (From Africa, always something new!)

Mystery areas: eastern Asia and Australia

The emergence of modern humans on the Asian mainland is poorly understood. Milford Wolpoff of the University of Michigan at Ann Arbor and his colleagues have argued that there was an indigenous evolution here of *H. erectus* into *H. sapiens*, but as I have suggested, this is not only theoretically unlikely but it's not well supported by any fossil evidence. For example, in China poorly dated archaic fossils such as a cranium from Dali (which may be a local version of European *H. heidelbergensis*) differ distinctly not only from Zhoukoudian *H. erectus* but from later fossils such as crania from Liukiang (about 34,000 years old, maybe older) and from the Upper Cave at Zhoukoudian (around 20,000 years old) that represent fully modern humans. Nothing that could really be regarded as "transitional" between the two is known. Indeed, Christopher Stringer of the Natural History Museum, London, has found mathematically that linking up these forms in a time series creates considerable problems. "It involves a more complicated change," says Stringer, "to go from Zhoukoudian to modern Chinese via Dali than to go directly." Moreover, he is unable to find any particular evidence of affinity between the Upper Cave and Liukiang specimens and today's Chinese.

It doesn't help, of course, that while new finds are being reported regularly in China they tend to be described scientifically with excruciating slowness. And proper description, of course, is essential before we can understand the significance of what are clearly some very important fossils. If this sounds like a paleontologist's lament, it is: once a fossil has been found you can't ignore it, but if you don't know what it looks like you can't include it!

In 1958 an expedition led by Tom Harrisson, the Curator of the Sarawak Museum in Kuching, western Borneo, excavated that island's spectacular Niah cave, a favorite spot for collecting the swifts' nests, plastered against the rock walls, that go into birds' nest soup. Among archaeological remnants the team found the skull of a teenager. Arguably, this skull is as much as 40,000 years old, which makes it the earliest modern human fossil known from the southeast Asian region. Since at that time Borneo would have been connected to mainland Asia by lowered sea levels, finding this early modern human there is no real surprise.

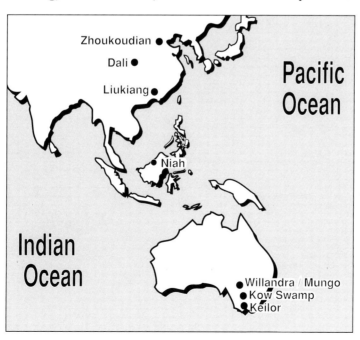

Various localities in southeastern Asia and Australia from which early modern human fossils have been recovered.

Illustration by Diana Salles.

Fragmentary crania of two ancient Australians. Right: Mungo 3, representing the earlier, "gracile", group. Left: Kow Swamp 5, a representative of the later "robust" group.

Photo by Willard Whitson.

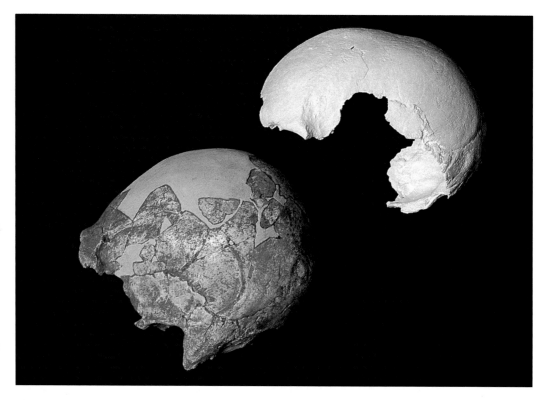

View of the dunes around the dried-up Lake Mungo, in southern Australia, where early Australians buried and possibly cremated their dead as early as 26,000 years ago. People first arrived in Australia well over 40,000 years ago and must have crossed at least sixty miles of open ocean to do so.

Photo by Dragi Markovic, courtesy of Alan Thorne.

Meadowcroft Rock Shelter, Pennsylvania: excavation in progress.

Courtesy of J.M. Adovasio.

View of Fells Cave, a site of early human habitation in Chile, near the southern tip of South America. As early as 11,000 years ago this remote place was inhabited by hunting peoples.

Photo by Junius Bird.

Alaska and Siberia were connected by a broad land bridge during times of lowered sea level. The white area here represents the emergent landmass of "Beringia" as it would have been about 20,000 years ago, at the height of the last glaciation; the heavy black lines show current coastlines, and the darkest areas represent ice sheets. Presumably following migrating herds of reindeer across low-lying Beringia, the earliest Americans must have followed an ice-free "corridor" to reach the ice-free southern areas of the North American continent.

Illustration by Diane Salles.

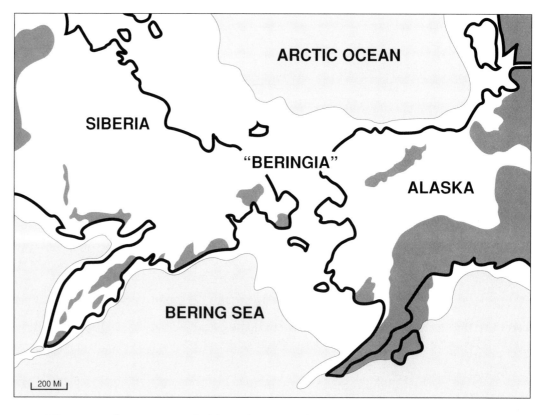

What is much more remarkable is that Australia appears already to have been inhabited by modern humans at this time. For even when the sea level was at its lowest (and at 50,000–40,000 years ago it was up—or down—to four hundred feet below its present level), the single landmass formed by Australia and New Guinea was separated from the Asian continent by at least sixty miles of ocean, and seaworthy boats and navigational skills would have been necessary to reach it. Such things figure, of course, among the many traits that distinguish behaviorally modern humans from all of their predecessors, and they speak of a remarkable degree of technological sophistication among the ancestral Australians.

We have no definite fossil evidence of the very earliest Australians, but a robust skull found by the paleoanthropologist Alan Thorne at Willandra Lakes in southeastern Australia may possibly be as much as 50,000 years old. This specimen (whose robustness may be due to disease, though this remains speculation at present) contrasts strongly with a group of human skeletons from sites such as Keilor and Lake Mungo, neither too far away, that date in the period of around 35,000–14,000 years ago and are of quite delicate build. These in turn contrast with younger specimens from sites in the same general region such as the one bearing the typically earthy Australian name of Kow Swamp. Although the Kow Swamp people may be as little as 9,000 years old they are decidedly robust, with quite noticeably projecting faces that project in front of braincases with receding foreheads.

The significance of this apparent recent reversion to a less modern-looking skull type is not very clear, especially since not many sites are known that have yielded the remains of ancient Australians. Various theories have been proposed, most of which involve the invasion of Australia on multiple occasions by people of different types. Another idea, though, is that at least in part the apparently archaic features of the more recent early Australians are due to artificial head deformation. The debate continues.

Late arrivals: the first Americans

Human populations appeared in the New World rather late compared to most other parts of the world, though the effects of this arrival were no less dramatic for that. Nobody knows when the first human set foot on American soil, or even exactly where, although these first immigrants were certainly Asian: "Biology, language, and archaeology," says the American Museum of Natural History's David Hurst Thomas, "all point to an Asian homeland." Almost certainly the first Americans entered the New World from Siberia across a land bridge exposed by lowered sea levels in the region where the Bering Straits are today; but the shallowness of the sea that now separates Alaska from Siberia indicates that the "bridge" would have been a broad one indeed, well over a thousand miles wide at its maximum. This huge emergent landmass would have provided vast grazing areas for Arctic mammals such as reindeer, and it seems likely that these early Siberians—the first humans culturally equipped to withstand the rigors of the Arctic climate—may have followed the migrating herds across into the new land.

Archaeologists energetically debate when this population movement first began. It has been claimed that humans were already present in both North and South America more than 50,000 years ago, but all claims for such an early origin for Native Americans are contested. Given what we know of the arrival of modern humans elsewhere in the world, and of the timing of lowered sea levels, something in the order of 30,000–25,000 years ago might seem a reasonable estimate; but there is no firm archaeological evidence until much more recently than that.

One of the more suggestive very early sites is Meadowcroft Shelter in western Pennsylvania, excavated during the 1970s and 1980s by a team led by the archaeologist J.M. Adovasio of Mercyhurst College. At this site a series of archaeological deposits has been radiocarbon dated to between 19,000 and 14,000 years ago, although the first stone implements turn up only in levels that are late in this period. The finds at Meadowcroft raise some questions, however. For example, the rather few stone tools are suspiciously similar to implements typical of a much later date, and the pollen preserved in the deposits came from plants that prefer much more temperate conditions than those that surely prevailed when the ice sheets were piled up a mere fifty miles to the north. Still, the site is an important and provocative one.

In any event, for unassailable proof of early human occupation of the Americas we have to wait until about 12,000 years ago. At this time evidence suddenly appears for the Clovis culture, named after a site in New Mexico. Clovis sites are known throughout the vast area of the American interior from Alaska to Guatemala, and it's obvious that the Clovis people were highly and rapidly successful. Perhaps the main reason for this was their remarkable ability to exploit the richness of the living resources that the New World had to offer. Most Clovis sites are places where woolly mammoths had been slaughtered; and indeed, such early Americans have been implicated in the widespread extinctions of large-bodied mammals that took place on this continent at the end of the last Ice Age. "Pleistocene overkill," the University of Arizona's Paul Martin calls it.

What's most remarkable about the Clovis is that there is no convincing evidence anywhere in the Americas—or in Asia, for that matter—for a culture ancestral to it. Archaeologists know a great deal about the development of the splendid diversity of Native American cultures in post-Clovis times: sites are abundant almost everywhere. But despite diligent searching, sites that document pre-Clovis human activities in the New World remain rare, and none is undisputed. Nonetheless, many archaeologists are convinced that the great variety in the kinds of stone weapon points found around the Americas right after 12,000 years ago means that the cultures that made them were diversifying for a long time, certainly thousands of years. But while cultures themselves disappear, the stone tools that they make are virtually indestructible. If people were around in the New World substantially before Clovis times, where are the sites that bear witness to their ancient activities? Archaeologists are still looking, and future finds may hold the key to the past; but meanwhile, the age and identity of the first Americans remains one of the great mysteries of the spread of humankind.

Where did we come from?

It is odd that the most recent major event in human evolution—the emergence of our own species—is perhaps the most obscure of all. The earliest well-dated modern humans are those fragmentary 120,000-year-old leftovers of a cannibal feast found at Klasies River Mouth in South Africa, and better dating may yet one day allow us to confirm that the very modern-looking skull from the Omo Basin is indeed 125,000 years old. For the time being, then, Africa seems to hold the age record for modern human fossils. But the Jebel Qafzeh people from Israel are almost 100,000 years old, and they look as modern as you could wish. And the more or less modern-looking people of Skhūl may be even older.

In short, the fossil evidence is painfully hard to read as to where and when our species originated. Once again, it seems, the familiar paleontologists' lament that the fossil record is just not complete enough is more than justified. Continued efforts to find new fossils are critical. And perhaps as importantly, we need to reappraise how the kinds of bony features that are preserved in fossils are distributed among modern human groups, as well as among the fossils themselves. This vital process will involve rethinking the entire way we look at the fossil record, and here the penetration of cladistics into

Two early modern human crania from the New World. On the left is the "La Brea Woman," some 9,000 years old, from the La Brea tar pits in Los Angeles, California; on the right is a reconstructed early cremation from the cave of Cerro Sota, near the southern tip of Chile.
Photo by Willard Whitson.

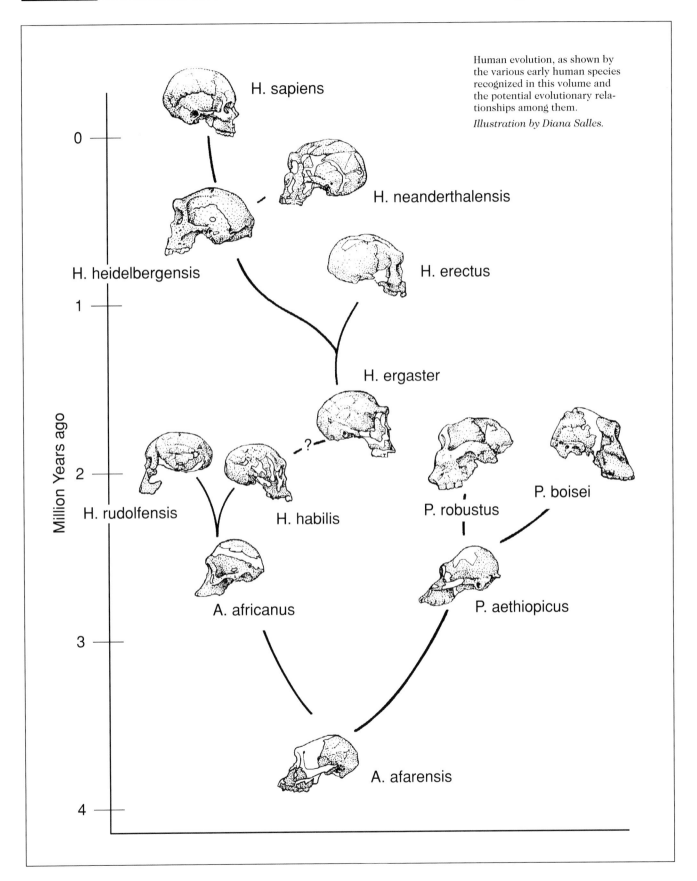

Human evolution, as shown by the various early human species recognized in this volume and the potential evolutionary relationships among them.

Illustration by Diana Salles.

paleoanthropology is already having a reinvigorating effect. Add to this the promise of gene-based technology for the anaysis of relationships, and the coming decades hold a lot to look forward to in unraveling the origin of our kind.

Yet people are not just anatomies or collections of molecules. When we seek to discover the origin of our species, we don't simply want to know when or where the first people who looked just like us strode the earth. What about the origins of the capacities and behaviors that make modern humans so unique? As we've seen, the archaeological record tells us that our precursors looked like us a considerable time before they started behaving like us. But when evidence of modern behaviors comes in, it comes in with a bang. We don't know where or when human creativity first arose; the record simply doesn't exist. But what we do know is that the first modern people who arrived in Europe left behind them proof of a creative spirit unprecedented in the entire 3-billion-year record of life on Earth. If there ever was a "human revolution," this was it. We'll look at this extraordinary record in the next chapter.

Chapter Twelve
The Human Spirit

Human cultures today exist in a dazzling variety. But a common thread runs through them all, based both on an extremely subtle understanding of the world that surrounds them and on an innate need to explain the place of humankind in that world. How far back in time can we trace that thread? A long way indeed, one might think, to account for the vast diversity of human cultural traditions; but the archaeological record shows that isn't so. For almost all of its 2.5-million-year span, that record reveals a story of highly episodic and strictly technological improvement—a new type of stone tool here, the introduction of hearths or rudimentary shelters there. Counting as one day the time since the first stone tool was made, it is only in the last twenty minutes that we begin to pick up archaeological evidence of the unique modern human sensibility, with its creativity, symbolism, and spirit of constant inquiry and innovation. But when we do, it is with a vengeance.

The part of the world that has provided us with this dramatic evidence is Europe, where the archaeological record of the last Ice Age is incomparably denser and more complete than anything known from elsewhere. This is hardly surprising when we consider how heavily this area is populated today, not least by archaeologists. There are, of course, hints of early behavioral modernity from other parts of the world: blade-based industries, for example, are known from Africa considerably earlier than in Europe, and the same continent has now begun to yield evidence for the decoration of objects as early as about 27,000 years ago, possibly earlier still. Evidence of the great age of prehistoric artistic achievements is also beginning to emerge from Australia. But since at this point early modern human attainments are so much better

Early twentieth-century photograph of an excavation at the important Upper Paleolithic site of Laugerie Basse, near Les Eyzies in southwestern France. The science of archaeology was low on its learning curve at this time.

Courtesy of the Department of Library Services, American Museum of Natural History.

View of the village of Les Eyzies, in southwestern France, the "capital of Prehistory." Along the valley of the Vézère River, in the vicinity of Les Eyzies, there occurs an incomparable concentration of Paleolithic sites, including many of the most famous of the Upper Paleolithic decorated caves. The famous rock shelter of Cro-Magnon is situated under the cliff on the right at the far end of the village.

Photo by Ian Tattersall.

The rock shelter of Cro-Magnon, in the village of Les Eyzies in southwestern France. In 1868 laborers building the nearby railroad discovered fossil bones here, following which the paleontologist Eduard Lartet and the banker Henry Christy excavated the skeletons of four adult modern humans and an infant. These skeletons were contemporary with bones of woolly mammoths, bison and reindeer, indicating a cold climate, and with tools and pierced seashells typical of the Aurignacian, the earliest period of the Upper Paleolithic. Rock shelters such as this were favored camping places during this time.

Photo by Ian Tattersall.

documented from Europe than from anywhere else, we'll look at this region as an example of what was undoubtedly happening more widely in the world during the same period. That period falls between approximately 35,000 and 10,000 years ago in the western and central European region where the record is strongest. Archaeologists know this period as the Upper Paleolithic, the last part of the Old Stone Age.

Cro-Magnon lifestyles

First, let's look at the way of life of the early modern Europeans who left us this record. They were the first people of modern appearance to inhabit this region of the world, and they are known as Cro-Magnons, after a site in southwestern France. In 1868 laborers constructing the railroad that passes through the small town of Les Eyzies dug fill from the small rock shelter of Cro-Magnon. In the rubble they found not only the skeletons of several *Homo sapiens* individuals who had been buried there, but also the remains of extinct animals that proved the antiquity of these burials, now estimated at about 25,000–30,000 years old. This was the first undoubted evidence of modern humans associated with ancient animals, and the Cro-Magnon people rapidly gave their name to all of the early inhabitants of Europe.

The laborers who dug the Cro-Magnon rock shelter did not leave a lot of evidence behind to allow archaeologists to reconstruct the lifeways of these people, but the discovery of many sites of similar and younger age just before and in the years following the Cro-Magnon find have provided us with an unparalleled knowledge of the lives of the early modern Europeans. Perhaps the best way of showing how totally these people represented a break with the past is to compare them with the Neanderthals, also well known from sites in this region and elsewhere.

Recent analyses of Neanderthal sites have suggested that their occupants may have scavenged the carcasses left by predators as often as they hunted for themselves. They

Reconstruction of the Magdalenian reindeer-hunters' camp at Pincevent, northern France, dated to about 12,000 years ago. Tepee-like structures of converging stakes covered with hides seem to have been quite typical of open-air encampments of this time.

Photo by Ian Tattersall.

built shelters rather infrequently. They most often made their stone tools from whatever stone was at hand rather than from the best materials for the job, maybe suggesting that they were not strong on anticipating their needs. Bone and antler were rarely if ever used as the raw materials for tools. There is a virtually complete lack of stone hearths at Mousterian sites in Europe, which hints that these Neanderthals used fire in a rather rudimentary way. Further, if we are to judge by the limited range of prey species among the animal bones at Neanderthal sites, their exploitation of the resources offered by the environment was rather limited. Evidence for decoration or elaboration of objects is virtually absent (except maybe at a couple of very late, Châtelperronian, sites—and as we've seen, this culture may have resulted from the Neanderthals' copying of innovations brought in by modern humans). There is virtually no evidence for any symbolic behavior (unless we can infer this from the very indirect evidence furnished by occasional burials of the dead), and none whatever for any form of notation. Of course, archaeologists are not unanimous on all of these points, but no one disagrees that the Cro-Magnons outshone the Neanderthals by a country mile in all of these categories.

The Cro-Magnons were without any doubt skilled hunters of game of all sizes, and they showed an intimate understanding of the environment in which they lived. They exploited this environment to the limit: fish and bird bones start appearing at Cro-Magnon sites where before they had been virtually absent from the archaeological record, and it's clear that these people regularly exploited the migratory movements of other vertebrates to their advantage. Campsites were often quite elaborate, and the making of complex fire hearths and the use of heated stones to heat up water in skin-lined pits shows that cooking had become a much more ambitious business. Cro-Magnons made tools out of bone, antler, and ivory, as well as out of stone and wood. They clearly understood the mechanical properties of each of these materials intimately; they even figured out that flint became easier to flake if they heated it up before they worked it. They

---

Behavioral contrasts between Cro-Magnons and Neanderthals. Cro-Magnon behavior represented a quantum break with anything previously seen in Europe. Just how dramatic this difference was is clearly shown by the comparison below of some behaviors reflected in the Aurignacian culture of the first Cro-Magnon people, as compared with the Mousterian culture of the Neanderthals.

	Mousterian	Aurignacian
Built shelters	x	x
Buried dead	x	x
Buried dead elaborately		x
Bodily ornamentation		x
Multiple materials used for tools		x
Subtle toolmaking methods		x
Use of materials from distant sources		x
Complex hunting methods; fishing		x
Complex use of fire		x
Decoration of objects; art		x
Musical instruments		x
Notation; symbolism		x

traded materials of decorative or utilitarian value over vast distances; for example, amber from the Baltic has been found at Upper Paleolithic sites in southern Europe, and Mediterranean seashells have turned up at sites in the Ukraine.

Cro-Magnons hafted stone flakes, and by at least 26,000 years ago the invention of eyed bone needles shows that carefully tailored clothing was being made. Material cultures began to vary from place to place, suggesting that local traditions were beginning to develop. Maybe these developments even included the evolution of different dialects or languages, for there is no question that these people possessed articulate speech. On the other hand, there's a lot of debate about whether the Neanderthals were capable of speech in the modern sense; a Neanderthal hyoid (throat) bone from the Israeli site of Kebara looks like its modern counterpart, but Jeffrey Laitman of New York's Mount Sinai School of Medicine points out that in the living individual much of the hyoid actually consists of cartilage, which the fossil doesn't preserve. "What's more," says Laitman, "the skull bases of the later western European Neanderthals are rather flat. This means that the voicebox was probably high in the Neanderthal throat, eliminating the long tubing above it that allows you and me to produce the full range of sounds involved in modern speech." This, he notes, presents something of a puzzle, because the voicebox was a bit lower in *Homo heidelbergensis*, the earlier species which may have been ancestral to us and to the Neanderthals.

As we saw earlier, the Neanderthals buried their dead at least occasionally, but among the Cro-Magnons burial became a regular occurrence. Some burials were more elaborate than others, and at times they were very elaborate indeed. One particularly striking case was uncovered at the site of Sungir, near Moscow. This site, excavated by Russian scientists during the 1960s and 1970s, is dated to about 28,000 years ago. An adult male and two children buried there had been dressed in clothing onto which literally thousands of mammoth tusk beads had been sewn. The laborious making of these beads must have involved at least as many thousands of hours of labor. Interestingly, the body of the older man buried at Sungir had been decorated in a way very similar to that of a young girl found far away at the much later site of La Madeleine, in southwestern France: bead bracelets ringed his wrists, elbows, knees, and ankles, and a band had encircled his head. The approximately contemporaneous Cro-Magnon rock shelter burials had apparently also been adorned with ornaments, in this case shell necklaces.

These and other similar finds suggest two things about the Cro-Magnon peoples. One was their love of bodily ornamentation, unprecedented among earlier humans unless the deposits of ochre found at Neanderthal sites had been intended for body coloration. The other is that those individuals who had been buried in such finery had enjoyed higher social status than others who were buried more simply. This in turn suggests that some degree of a social stratification existed, which hints at considerable complexity in Cro-Magnon societies. Certainly there was some division of labor: at numerous sites, vast piles of chipped stone reveal the existence of "workshops" at which stone tools were made, presumably by specialists. Whether these stoneworkers enjoyed high status in their societies we'll probably never know.

Cro-Magnons also devised various ingenious ways of exploiting their environments to the fullest. For example, by about 15,000 years ago the inhabitants of the Eastern European Plain had developed a method of preserving meat in pits dug into the permafrost. How they coped with the problem of freezer burn is unknown, but this invention allowed these people to live more or less permanently in a single locality, even at times of year when the reindeer herds had moved away.

Diorama at the American Museum of Natural History showing one of the mammoth-bone huts excavated at the 15,000-year-old site of Mezhirich, in the Ukraine. This hut is one of four, each of which had a distinctive arrangement of mammoth bones covering its sides; for this reason, this ancient village is sometimes regarded as exhibiting the earliest architecture. This hut was made from some seventeen tons of bones representing the remains of ninety-five mammoths, which presumably had died from natural causes on the surrounding steppes.

Photo by Dennis Finnin.

Naturally, not all of these and other innovations were introduced instantaneously; a process of refinement went on throughout the Upper Paleolithic. But without question, altogether modern behavior patterns were established by the time the Cro-Magnons arrived in Europe. In nothing is this so evident as in the case of art and symbolism, which are perhaps the most vaunted of our own accomplishments. In the rest of this volume we'll look at the extraordinary artistic and symbolic achievements of the cultures of the late Ice Age.

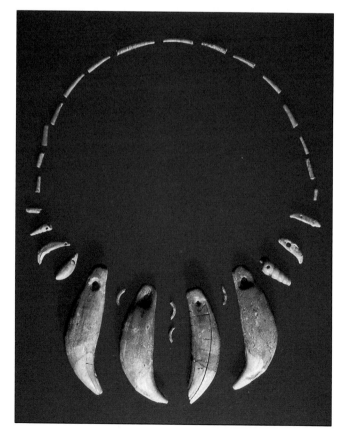

Pierced shells and teeth, reconstructed as a necklace. The three largest teeth are the canines of cave bears, already extinct for almost 40,000 years when these were collected (presumably from skulls lying around in caves) and pierced. The fourth large tooth is that of a lion. These objects are from the late Magdalenian site of Rocher de la Peine, France, about 12,000 years old.

Photo courtesy of R. White.

Ice Age art

As we've seen, early representational art is not known solely from Europe, but it is certainly known best from this region, where it appeared alongside the earliest modern humans. Unprecedented though it was, much of this early art was not in the least crude or primitive. From the very beginning, the finest Paleolithic art showed a perceptiveness of observation and a command of form that rivals anything achieved since. Animals are the most familiar subjects portrayed by Ice Age artists, but both human and mythical representations were made as well, and a wealth of geometric and abstract designs is known. Both monochrome and polychrome paintings were made on cave walls, using naturally occurring pigments; bas-reliefs were carved on the walls of rock shelters; clay figures were molded and sometimes fired in kilns; stone and bone plaques were engraved; animal and human figures were carved in antler, ivory, bone, and stone; and numerous utilitarian items were decorated by carving and engraving.

So acutely observed that we can sometimes tell both by behavior and appearance in what season of the year an animal was depicted, the art of these Ice Age people had profound symbolic significance: although often very beautiful, it was never merely decorative. Rarely do we find animal figures that are not accompanied by abstract symbols of some kind. And although at breathtaking sites such as Lascaux in southwestern France coherent compositions of animal figures cascade across the walls of the cave, it is clear that such compositions are not simple representational scenes. Each animal clearly has

Image of an ibex (mountain goat), drawn deep underground on the wall of the cave of Cougnac, in southern France, some 14,000 years ago.

Photo by Alain Roussot.

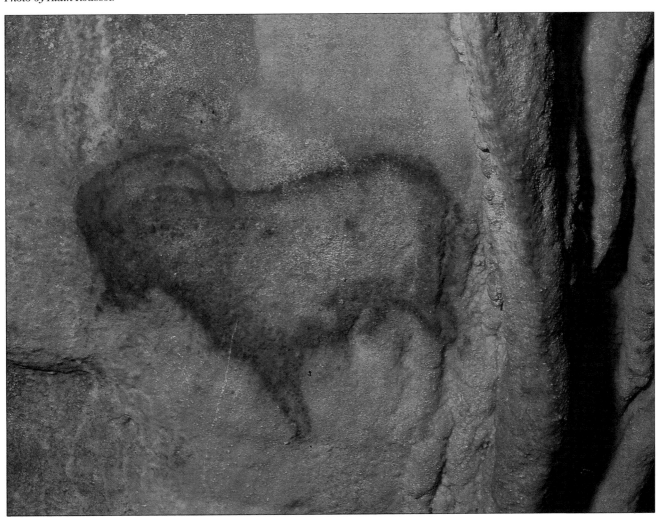

its place and its significance, but to the artists who painted these astonishing images the animal figures may have had no more (or less) symbolic significance than the abstract designs that accompany them. Not for nothing has the New York University archaeologist Randall White described Ice Age art as the "most widely known but least understood aspect of Upper Paleolithic life."

The symbolic value of the animal images is emphasized by the fact that some were used over and over again by Ice age artists: such images at several sites were overengraved or overpainted, or had extra details added, such as supplementary legs, ears, or tails. The artists who did this were probably not simply improving or improvising on the work of their forebears, however: such modifications seem often to have been made by artists of the same time period. For example, the cave of La Mouthe near Les Eyzies was decorated over a period of several thousand years. This tortuous, winding gallery penetrating deep into a limestone hillside is decorated along a length of several hundred feet, and the images inside it become progressively younger as one proceeds deeper into the interior. Apparently the Ice Age artists respected the work of their predecessors and did not wish to encroach on it; as a result, the earlier art "forced" later artists deeper into the cave. Perhaps we should thus see the modification of at least some Ice Age images as a sort of "performance art," in which the act of making or modifying the images was at least as important as the existence of the end result.

Another factor that suggests that the meaning of most of the Ice Age images that have survived on the walls of caves was essentially a ritual or symbolic one is their very inaccessibility. We can only guess at the exact motivations that impelled Upper Paleolithic artists to penetrate into pitch-dark, dangerous and uncomfortable recesses deep within the earth by the light of flickering lamps made from juniper wicks stuck into lumps of animal fat, but we can be sure that they were profound and powerful indeed. Nonetheless, although to most of us today these places seem to be forbidding, unsettling environments—even when we visit them with the benefit of profesional guides and

Engraving on a pierced "baton" of reindeer antler, showing two stags and several fish; possibly it represents deer crossing a river during the salmon run. Although the design is shown here flattened out, it was actually executed on a cylindrical surface. This is from the Magdalenian site of Lortet, in the French Pyrenees.

Illustration by Don McGranaghan.

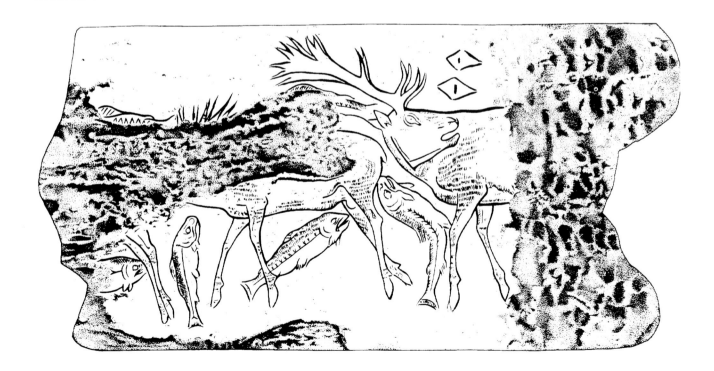

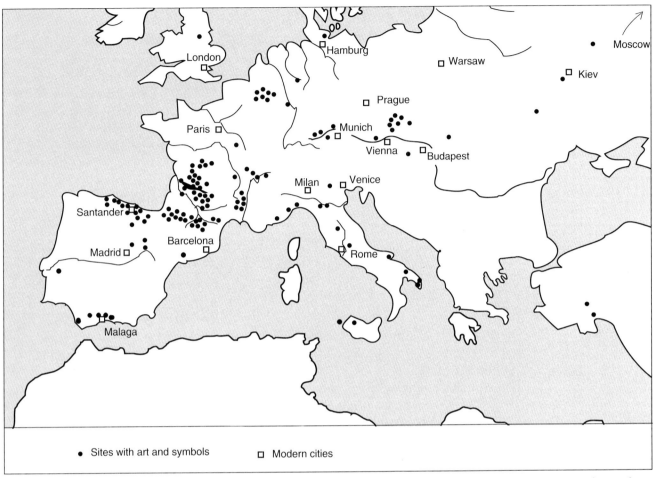

The principal sites (dark circles) at which Ice Age artworks have been found in Europe. Sites with cave wall art are concentrated in southern and southwestern France and northern Spain. Portable art is found much more widely.

Illustration by Diana Salles.

modern lighting, and although they seem to have been shunned by hundreds of generations of people living near them since the end of the Ice Age—it is not at all clear that they seemed that way to the people of the Upper Paleolithic.

The archaeologist Christine Desdemaines-Hugon points out that deep in one cave there are some tiny human finger-tracings in the clay covering the cave wall. "These," she says, "must have been made by a tiny child, and one can just imagine the hand of a parent or other adult guiding the child's fingers across the wall." In another cave tiny footprints, undisturbed over the millennia, testify to the presence of small children in those cavernous depths, hundreds of yards from the light of day, at least 14,000 years ago. If infants were taken along with the artists into these pitch-black, difficult surroundings it is hard to conclude, as it is so tempting to do, that entering these places was somehow a test of will or endurance. Special these places undoubtedly were to the people of the Ice Age, but not necessarily because of their potentially terrifying nature.

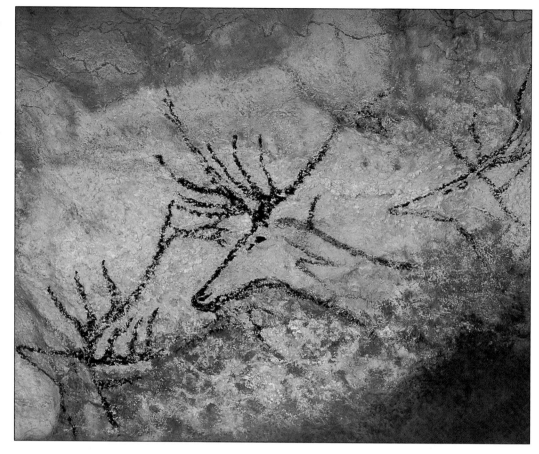

Scene painted on a cave wall at Lascaux, in southwestern France. This Magdalenian cave is renowned for its spectacular polychrome paintings of animals as well as for numerous delicate animal engravings and geometrical signs. The monochrome design seen here shows a series of deer heads, interpreted by some as representing a single deer plunging into and swimming across a river. If so, the water's surface is represented by the natural contours of the rock wall. This scene is about 17,000 years old.

Photo by Jackie Beckett

The development of Ice Age art

Until recently, when advances in radiocarbon technology have allowed tiny samples of organic pigments such as charcoal to be dated in years, there was no direct method available of dating any of the Ice Age cave images. Proof of their ancient age was furnished by heavy deposits on them of calcite, a limestone film that forms very slowly on the cave walls, or by overlying archaeological deposits containing Upper Paleolithic tools. But generally all that could be said was that where images overlapped the one above was younger than the one below—though whether by minutes or by millennia remained moot. Since most such works were engraved, or were drawn or painted using inorganic pigments such as ochre or manganese, we are even now forced to rely for dating on stylistic comparisons of cave art with the "portable" art that was sculpted, molded, or engraved on bone, stone, or ivory plaques. These can be dated with confidence because they are found in stratified archaeological layers, in association with tools, animal bones, and other refuse.

Four successive phases of the Upper Paleolithic are commonly distinguished in western Europe. From oldest to youngest these are the Aurignacian, Gravettian, Solutrean, and Magdalenian. Technically, each of these periods is defined by a particular tool assemblage; but each, too, also showed its own characteristic forms of artistic expression. Small-scale decorated objects are ubiquitous throughout the Upper Paleolithic and from both eastern and western Europe, but cave-wall paintings and engravings have a more limited distribution. These, most strikingly of animals but also with many abstract symbols and occasional human or imaginary representations, are largely concentrated in the area of southwestern France and northern Spain, and in the period of about 20,000–

12,000 years ago. As we have seen, such images and the activities that were presumably associated with them were clearly of especial significance in the cultural traditions of the people of this time and place. Over this long period artistic styles changed, but it is debated whether this change represented a continuous long-term development over thousands of years, as the eminent French prehistorian André Leroi-Gourhan concluded it was.

The earliest examples of representational art known are elegant animal carvings from the German site of Vogelherd, dated at about 32,000 years ago and thus right around the beginning of the Aurignacian, which lasted in western Europe from something over 32,000 years ago until about 28,000 years ago. From about the same remote time the French site of Isturitz has produced a bone flute with a complex sound capability, and from the Abri Blanchard has come an engraved plaque with complex markings that the analyst Alexander Marshack has interpreted as a lunar calendar. Art, music and notation were thus not late developments in human history; instead, they embody the very essence of what it means to be behaviorally fully human. Besides animal sculptures and engravings in mammoth ivory and other materials, a surprising diversity of Aurignacian graphic forms included engraved images of sex organs and animals. In some cases, engraved blocks from this period retain traces of paint. Perforated stone, bone, and ivory beads and pendants provide evidence of bodily adornment.

Deeply incised engraving of a horse on a massive limestone block. The style is typical of the Gravettian period, with outwardly sloping edges to the incisions producing the illusion of bas-relief. Traces of red pigment remain on parts of the image. This is from Abri Labattut, near Sergeac, western France, and is probably about 25,000 years old.

Photo courtesy of the American Museum of Natural History.

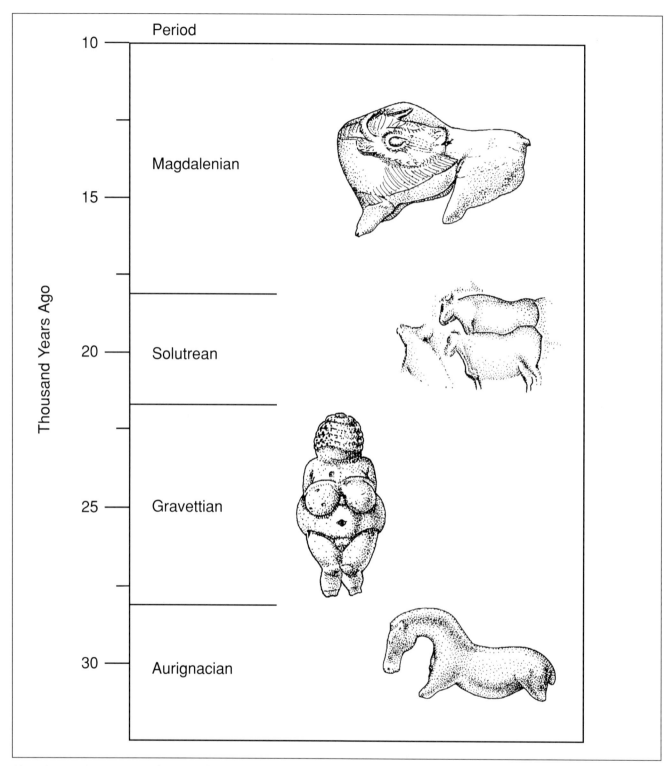

Chart showing the succession of periods of the Upper Paleolithic in western France. Each of these periods is defined on the basis of particular tool types, but each period also had its characteristic art forms, as seen on the right. In other areas of Europe, the succession is slightly different: the Solutrean is not recognized east of the Rhone Valley and is replaced by an "Epigravettian" derivative of the Gravettian culture, which notably in the east of Europe continued to overlap with the Magdalenian.

Illustration by Diana Salles.

More complex art forms emerged in the succeeding Gravettian period, from about 28,000 to 22,000 years ago. The most famous of these are the female statuettes and large engraved or sculpted female forms often known as "Venus" figures. Many of these, such as

One of the world's earliest artworks: a 32,000 year-old carving, in mammoth ivory, of a horse. This extraordinarily elegant sculpture, less than two inches long, does not literally represent the stocky proportions of Paleolithic horses. Rather, it conveys the graceful essence of the horse. This is from the early Aurignacian of Vogelherd, Germany.

Illustration by Don McGranaghan.

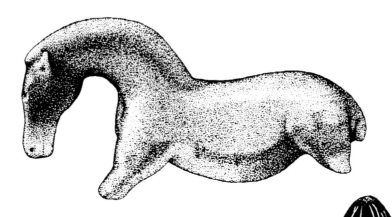

the Venus of Lespugue, show exaggerated breasts and buttocks, with de-emphasis of limbs and individual features; others, such as those from Brassempouy, are more modestly proportioned. Venuses are often interpreted as fertility symbols, but the fact is that lack of fertility is rarely a problem for nomadic hunting-gathering peoples, among whom needing to care for too many offspring while on the move can cause enormous difficulties. From Brassempouy also comes what may be a representation of an actual person, a rarity in Upper Paleolithic art.

Particularly interesting innovations in the Gravettian are molded clay figurines, both of human females and of a variety of mammals. Figurines of this kind have been recovered at the site of Dolni Vestonice in Czechoslovakia, and date from about 26,000 years ago. They were fired in simple kilns at up to 800 degrees Fahrenheit, and they represent the first recorded use of the techniques that were essential to the development of pottery some 16,000 years later. Like so many later innovations, the firing of clay was thus invented in a decorative, symbolic context rather than a practical one.

The "Venus of Lespugue." This figure, sculpted in mammoth ivory, is one of the most elaborate of the "Venus figurines" that are so characteristic of the Gravettian period. Typically, facial features are lacking, the breasts, abdomen and buttocks are emphasized, and the legs and feet are schematic. The figure was badly damaged during excavation. It is from Lespugue, southern France, and is about 23,000–24,000 years old.

Illustration by Diana Salles.

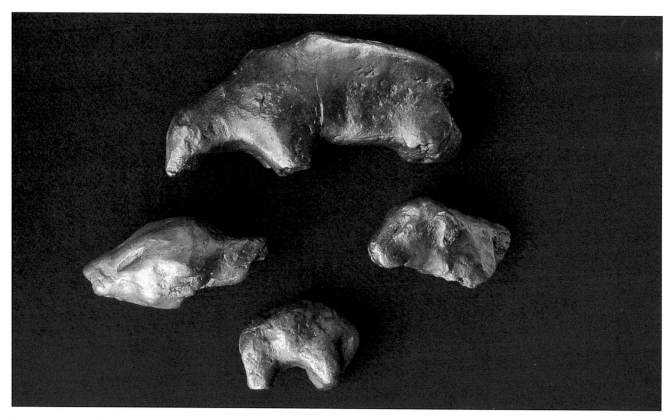

A bear, a mammoth, a bear head, and the head of another carnivore, all molded in clay and kiln-fired at high temperatures. They represent the Gravettian of Dolni Vestonice, Moravia, about 26,000 years ago.

Photo by Peter Siegel.

During the Solutrean period, between 22,000 and 18,000 years ago, the art of flint-working was brought to its highest level of excellence. Particularly characteristic of the period are the so-called "laurel-leaf" blades, up to a foot long and pointed at both ends. These were sometimes so thin and delicate as to suggest that they were purely of symbolic or ceremonial significance. The representational art of the Solutrean is characterized by large-scale bas-relief sculptures, often in the form of friezes of animal images on rock shelter walls adjacent to living areas. At this time the practice of painting in deep and inaccessible cave recesses also became well established, although the Gravettians also had done so occasionally. Typical Solutrean tools and images are actually recognized only in northern Spain and in France to the south of the Loire and east of the Rhône rivers; in other areas of Europe we generally find continuations in one form or another of the Gravettian tradition. Some of these were very long-lasting indeed, overlapping considerably with the next western European phase.

The final period of the European Upper Paleolithic, the Magdalenian (18,000 to 10,000 years ago), witnessed an astonishing richness and diversity of artistic achievement, based on an acute observation of nature and a total mastery of materials. Magdalenian artists "conquered" the underground, penetrating remote cave chambers to execute some of their greatest works: the caves of Lascaux and Altamira, among many others, attest eloquently to their command of complex color painting. Bas-relief friezes continued to appear on rock walls, and innumerable small objects, both tools and plaques, were elegantly carved and engraved. The intricate associations in Magdalenian art between different animal species, and between animal representations and abstract symbols, probably reflect a complex body of myth, story, and belief. Once again, most of the intricate compositions of animal and geometric designs defy any literal interpretation of what the whole represents, but a classic example of what must be a mythical scene

Scene from the "well," a ten-foot-deep cavity in Lascaux cave, southwestern France. This is the only representation of a human at Lascaux and, unlike all of the many animal images painted elsewhere in the cave, is highly schematic. The bird-headed and ithyphallic human falls before a bison that has apparently been disemboweled by a lance, while a rhinoceros retreats to the left; and the scene is completed by a variety of abstract or semi-realistic symbols such as the bird atop a pole. We may well see a myth recounted here, but a myth replete with many layers of meaning. This was drawn about 17,000 years ago.

Photo by Dr. N. Aujoulat, Département d'Art Pariétal, Centre National de Préhistoire.

appears deep in a pit in the cave of Lascaux. Here a falling (dying?) man is attacked by a bison which appears to be disemboweled by a spear that pierces its flank. Behind him a woolly rhinoceros retreats, and below him is a bird sitting atop a pole. Interestingly, at the cave of Villars, some thirty miles to the northwest of Lascaux, there is another depiction of a man confronting a bison. If this represents a simplified version of the Lascaux scene, we may be seeing here a visual recounting of a story current in the oral tradition of the time.

What that story was is anybody's guess. A couple of decades ago the French archaeologist François Bordes, stepping outside his strictly scientific role, was brave enough to hazard one. "Once upon a time," he surmised, "a hunter who belonged to the bird clan was killed by a bison. One of his companions, a member of the rhinoceros clan, entered the cave and drew the scene of his friend's death—and of his revenge. The bison is pierced by weapons and is disemboweled, probably by the horns of the rhinoceros." An intriguing possibility, but one which places the scene in the context of a particular incident rather than in the realm of mythology, which is where its possible repetition at two sites suggests may be more appropriate.

But Bordes's explanation—not a serious one, of course—does echo one of the earliest general explanations of Ice Age cave art, which was that of sympathetic magic, or the idea that symbolic actions may impact real life. Just as the bloated "Venus" figurines were interpreted as possible fertility figures, the representations of animals on cave walls were thought by many, including the Abbé Breuil (1877–1961) who dominated the study of cave art during the first half of this century, to represent hunting magic. Some of the animal—and human—images on the cave walls are penetrated by lines, thought by many to represent spears. Under the sympathetic magic explanation, the symbolic killing of animals on cave walls might have been aimed at assuring success in the hunt.

Such explanations, though, ignore one important observation. Archaeological remains show that by far the most commonly hunted animals in the late Ice Age were reindeer. And reindeer are among the least represented large mammals in Ice Age art. What's more, the most famous reindeer depiction in Ice Age art is hardly the kind of image one associates with hunting and killing. In the cave of Font de Gaume, just outside Les Eyzies, there is a badly faded but immensely moving painting of a pair of reindeer. On the right is a female, bending down with forelegs flexed, facing a male with magnificent antlers. In his turn, the male is leaning forward and gently licking the female's forehead. The scene is beautifully observed, and full of tenderness. I, for one, am totally incapable of seeing this image in any violent context.

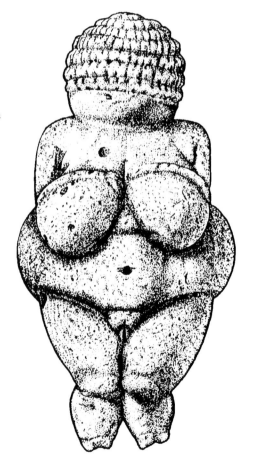

The "Venus of Willendorf." This figure is perhaps the most famous of all of the "Venus figurines" that typified the Gravettian period. Sculpted in limestone, it is typical in showing no facial detail, emphasizing the breasts, buttocks and abdomen, and minimizing the lower limbs. From Willendorf, Austria, it is about 25,000 years old.

Illustration by Diana Salles.

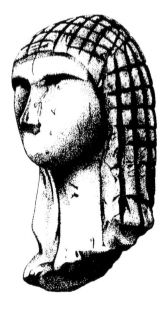

The "Lady of Brassempouy." One of the few Gravettian female sculptures that may represent an actual person, this image is engraved in mammoth ivory with extreme delicacy. From Grotte du Pape, Brassempouy, southwestern France, it is probably about 25,000 years old.

Illustration by Don McGranaghan.

Ice Age art and the environment

The greatest achievements of Ice Age art cluster in the millennia around and following the peak of the last glaciation, which occurred about 18,000 years ago. This may on the face of it seem surprising, but in reality cold times were not necessarily hard times. The open steppe vegetation that covered vast areas of Europe at this period sustained huge populations of large-bodied herbivorous mammals in an open habitat that provided relatively easy conditions for hunting. Alexander Marshack, a specialist in Ice Age symbolism, also has pointed out that the area of northeastern Spain and southwestern France where Ice Age art was at its richest, was greatly favored by geography. This region was relatively sheltered, with a varied landscape that provided a wide range of habitats, from the hilly crags inhabited by the agile ibex right down to the valley floors where the rivers teemed seasonally with migrating salmon. The basic business of making a living was probably not too time-consuming in so productive an environment, and the diverse landscape and fauna would in themselves also have offered a wide range of stimuli to the symbolic and artistic imaginations of people who had all of the intellectual equipment that we have. Much of the Cro-Magnons' time must thus have been available for cultural pursuits in the widest sense, and this helps explain, for instance, the enormous numbers of decorated objects found at some sites.

What this art meant to these people, members of long-vanished societies, we will of course never know. There is a basic human urge to decorate and to elaborate, and doubtless much of the carving and engraving of small objects was done in the service of this essential human instinct. But the motivations of the cave artists, in particular, and of the societies that supported them (for some, at least, must have been professionals), were much more complex and mysterious than that. In some fashion this art reflected the way in which these societies understood the world around them and explained it to themselves, and it's highly unlikely that we'll ever know exactly what it meant to them. From our modern perspective, then, this will remain art to be experienced with awe. It is not art to be explained by us, however much it is a basic human instinct to want to explain everything. Anyone who has the privilege of seeing these magnificent echoes of the past will, however, surely agree that experience is enough.

Monochrome rendering of a badly faded polychrome wall painting in the cave of Font de Gaume, France. A female reindeer kneels before a male, who leans forward and delicately licks her forehead. This was probably painted about 14,000 years ago.

Illustration by Diana Salles, after a rendering by the Abbé Breuil.

Delicate engraving of a horse on a bone plaque. This image dates from about 9,600 years ago and is thus from the Azilian epoch, which followed the Magdalenian. This graceful animal image is somewhat unusual for this "Epipaleolithic" culture which was materially much less rich than the Magdalenian. It is from Pont d'Ambon, France.

Photo courtesy of Jean-Jacques Cleyet-Merle.

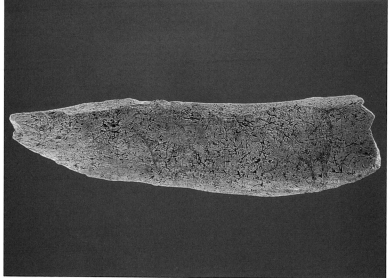

Delicate flint "laurel leaves," the longer of them a foot in length and typical of the Solutrean period. Many of these elegant blades were so thin and delicately crafted that their use must have been purely ceremonial or symbolic. These are from Volgu, France and are about 20,000 years old.

Photo by Peter Siegel.

Profile of a bison, subtly sculpted in bas-relief on a cylindrical baton of reindeer antler. This work is from Isturitz, France and is about 14,000 years old.

Illustration by Don McGranaghan.

The end of the Ice Age

About 12,000 years ago a warming trend began to usher in the replacement of European steppe/grasslands by forests. As the last glaciation came to an end, many of the large herbivores that had sustained the Magdalenian hunters became extinct, and the populations of others were dramatically reduced. Hunting red deer and boar in the newly widespread forests was a very different—and much more difficult—proposition from hunting herds of reindeer and bison on the open steppes. These ecological changes thus represented a disastrous decline in the resources that had supported the Magdalenian hunters. With this economic deterioration came social change, and the great animal art of the Upper Paleolithic disappeared by about 10,000 years ago. Geometric and abstract images became the dominant forms of artistic expression in the relatively impoverished new social and economic milieu. In a rather short time Europe became a cultural backwater, and the focus of technological and social innovation shifted toward the east, where the era of settled agriculture was about to dawn.

Although the hunters of the Ice Age were masters of their environment to an extent achieved by no people before them, they and their high cultures thus ultimately proved to be as much the prisoners as the beneficiaries of that environment. We may never fully understand their art, but there is a message in their fate for today's humanity. For we, in our turn, pride ourselves on our mastery of an environment which in all likelihood is more fragile and changeable than we may like to suppose.

Glossary

Acheulean

Lower Paleolithic stone toolmaking tradition especially noted for large, bifacially flaked hand axes and cleavers. It first appeared about 1.5 million years ago in Africa and lingered in Asia until 150,000 years ago or less.

African Eve

African common ancestor of all modern human populations, as inferred in certain studies of mtDNA variation among living human groups. It is called "Eve" because all mtDNA is inherited from the female parent.

Altricial mammals

Mammals in which litters of young are born at a relatively undeveloped stage, following a short gestation period. Parental investment in each offspring tends to be low.

Amino acids

Nitrogen-containing molecules that are the "building blocks" from which proteins are made.

Antibodies

Proteins produced by white blood cells in response to attack by foreign substances.

Aurignacian

First phase of the Upper Paleolithic in Europe. It began about 40,000 years ago in Eastern Europe and lasted until about 28,000 years ago.

Autonomic nervous system

The part of the nervous system that is concerned with involuntary activities: breathing, digestion, etc. It is subdivided into the opposing sympathetic (stimulating) and parasympathetic (inhibitory) divisions.

Axon

A long branch of a nerve cell that conducts nervous signals outward, away from the cell body.

Bacteria

Organisms, also called prokaryotes, that lack a membrane-bound nucleus.

Mural painting, executed by Charles R. Knight for the American Museum of Natural History in 1920, showing the decoration of a Paleolithic cave by a group of Cro-Magnons. Although it reflects a rather archaic understanding of the Cro-Magnons themselves, it nonetheless evokes the strange and wonderful atmosphere of these remarkable places.

Courtesy of the Department of Library Services, American Museum of Natural History.

Bases

The "rungs" of the DNA "ladder." DNA bases are of four kinds: adenine (A), thymine (T), guanine (G), and cytosine (C). A and T always pair together, as do G and C.

Beringia

Name given to the land area that linked Siberia and Alaska at times of lowered sea levels (during glacial periods).

Blade

An elongated flake, more than twice as long as wide, struck from a prepared semicylindrical stone core. Blades may be "retouched" into a wide variety of different implements.

Brachiation

Arboreal arm-swinging locomotion in which the body is suspended from branches and propelled along by the forelimbs alone.

Breccia

A rock composed of more or less coarse and angular rock fragments cemented together by lime.

Carbon-14 dating

See radiocarbon dating.

Catarrhini

Infraorder of Primates that includes Old World monkeys, apes, and humans.

Châtelperronian

Stoneworking industry variously viewed as the last stage of the Middle Paleolithic or the earliest stage of the Upper Paleolithic in western Europe. Probably the work of Neanderthals, possibly after seeing the Upper Paleolithic technology of the earliest modern humans in the area. It dates to about 36,000–32,000 years ago.

Chronometric dating

Dating of fossils by assigning ages in years. Most methods depend on the steady rates of breakdown of unstable (radioactive) isotopes; most familiar among these are radiocarbon and potassium/argon dating.

Cladistics, cladism

An approach to the reconstruction of evolutionary relationships among species and higher taxa that depends on the recognition of shared derived features.

Core

Lump of stone from which flakes are struck to produce tools.

Cretaceous

Final period of the Mesozoic era, the Age of Dinosaurs, about 145–65 million years ago.

Cro-Magnon

Name generally used for the earliest modern human inhabitants of Europe. It is derived from the rock shelter of Cro-Magnon in western France, where the remains of ancient anatomically modern humans were first discovered in association with the fossils of extinct animals. Cro-Magnon fossils are known as early as about 35,000 years ago, and the term ceases to be applied with the beginning of the Epipaleolithic about 10,000 years ago.

Dendrite

A branch of a nerve cell that conducts nerve impulses inward, toward the cell body.

Dentition

Teeth.

Derived

A characteristic that has departed from the ancestral condition.

Diurnal

Active during the daytime.

DNA

Deoxyribonucleic acid, the self-replicating molecule that transmits hereditary information across generations.

Electron spin resonance

Method of dating objects by determining the number of trapped electrons in a sample using their absorption of microwave radiation. Has been used on dental enamel.

Endotherm

Organism, such as a mammal or bird, that maintains a high and constant internal body temperature.

Eocene

Geological epoch, 58–35 million years ago.

Eukaryote

Organism possessing cells with membrane-bounded nuclei and other organelles.

Euprimates

"Primates of modern aspect," including all primates living today, and all extinct primates except the plesiadapiforms.

Evolution

Charles Darwin's original definition is still the best: "descent with modification."

Family

Rank in the hierarchy of animal classification that lies below the superfamily and above the subfamily, which in turn lies above the genus.

Faunal dating

Dating of rocks and the fossils they contain by comparing those fossils with those known from rocks elsewhere.

Flake

Fragment struck from a larger piece of stone, using either direct percussion or pressure. Such fragments may be "retouched" into tools of varying kinds.

Gene

The unit of heredity, composed of a specific length of DNA in an individual's genome.

Genome

Sum total of an individual's DNA.

Genus

Larger category into which related species are grouped, and the first component of a species name. For example, the species *Homo sapiens* belongs to the genus *Homo*, along with several other species, all now extinct.

Glacial

A cold period in which the polar and mountain ice caps spread as the accumulation of snow on them exceeds the rate of melting at their edges.

Gravettian

Second major Upper Paleolithic culture in Europe, well known for the production of "Venus" figurines. It lasted from about 28,000 to about 22,000 years ago in the western part of the continent. Rather similar Epigravettian traditions lasted much longer farther east.

Heterotroph

Organism that is unable to derive energy from photosynthesis or inorganic molecules, and thus must do so by feeding on other organisms.

"Higher" primates

Primates belonging to the suborder Anthropoidea, including monkeys, apes and humans, and their extinct relatives.

Holotype

The designated name-bearer of a species and the specimen to which all putative other members of the species must be compared.

Hominid

Member of the family Hominidae, nowadays reckoned by most to include the great apes as well as living humans and their fossil relatives. Earlier use restricted this family to living and fossil humans alone.

Hominoid

Member of the superfamily Hominoidea, which embraces humans, the great and lesser apes, and their fossil relatives.

Interglacial

Warmer period between glacial episodes, characterized by the retreat of the ice caps.

"Lower" primates

Primates belonging to the suborder Strepsirhini, including the lemurs, lorises, and bushbabies, plus their fossil relatives and possibly the tarsier.

Magdalenian

Final phase of the Upper Paleolithic in Europe, about 18,000–10,000 years ago, especially noted for its artistic achievements.

Marsupial

Pouched mammal.

Miocene

Geological epoch, about 24–5 million years ago.

Monotreme

Egg-laying mammal.

Mousterian

Flake-based stoneworking industry associated in western Europe with the Neanderthals, but also produced in the Near East by early populations of anatomically modern humans and best known from the last glaciation about 100,000–35,000 years ago.

mtDNA

Extranuclear DNA found in the mitochondria, the "powerhouses" of eukaryotic cells. All of an individual's mtDNA is inherited from the mother, since the ovum is a complete cell while the sperm contains only nuclear DNA.

Natural Selection

Individuals vary within every population, and those with certain inherited traits will reproduce more successfully than others. Over the generations such differential reproduction will change the genetic makeup of the population: this pressure for change is known as natural selection.

Neanderthal

Name given to members of *Homo neanderthalensis*, an extinct human species that inhabited Europe and western Asia about 150–35,000 years ago.

Neuron

Nerve cell.

New World Monkeys

Primates of the infraorder Platyrrhini, the primates of South America, including marmosets, tamarins, capuchin, howler, and spider monkeys.

Nocturnal

Active at night.

Oldowan

Earliest known stoneworking tradition, which appeared in Africa about 2.5 million years ago and in which small simple flakes were detached from small cores.

Old World Monkeys

Monkeys belonging to the superfamily Cercopithecoidea of the infraorder Catarrhini, found in Africa and Asia, including leaf monkeys, proboscis monkeys, baboons, macaques, and vervets.

Oligocene

Geological epoch, about 35–24 million years ago.

Paleocene

Geological epoch, about 65–58 million years ago.

Paleolithic

The Old Stone Age, which began in Africa about 2.5 million years ago and was supplanted only quite recently. It was first supplanted in the Near East about 10,000 years ago, with the introduction of agriculture and the polishing of stone. The Oldowan and Acheulean industries are regarded as Lower Paleolithic, while the succeeding Middle Paleolithic and Upper Paleolithic periods of Europe are roughly equivalent to the Middle and Late Stone Ages of Africa.

Parasympathetic nervous system

See autonomic nervous system.

Photosynthesis

Process whereby plants synthesize organic compounds using the energy of sunlight.

Phylogeny

Evolutionary history of a group.

Placental mammals

Mammals in which the young develop to an advanced stage within a mother's body before birth, nourished from the mother's bloodstream via the placenta.

Platyrrhini

Infraorder of Primates that includes monkeys of the New World.

Pleistocene

Geological epochs, popularly known as the Ice Ages, that began about 1.6 million years ago. Its ending date, generally considered to be about 10,000 years ago, may not have arrived yet.

Plesiadapiforms

Group of archaic primates that came into existence right at the beginning of the Age of Mammals, and became extinct in the early Eocene. They do not closely resemble any primates living today.

Pliocene

Geological epoch, about 5–1.6 million years ago.

Precocial mammals

Mammals in which small numbers of young are born at a highly advanced developmental stage following a relatively long gestation period. Parents usually invest considerable energy in each offspring.

Primates

The order of mammals to which humans belong. The first letter is capitalized when the word is used as a proper noun; not otherwise.

Prokaryote

Literally "pre-nucleus," a bacterium, lacking the organized cell structure that characterizes other life forms.

Protoctist

Member of the kingdom of living organisms that groups together such miscellaneous eukaryotes as seaweeds and slime molds, which are neither fungi, plants, nor animals.

Quadruped

Tetrapod that uses all four limbs to move around.

Radiation (adaptive)

Diversification of a number of species (or higher taxa) from a common ancestral species.

Radiocarbon (carbon-14) dating

Method of obtaining year-dates in organic materials by measuring the ratio of unstable (radioactive) carbon in a sample to stable carbon. It can sometimes be used directly on bone and is useful to about 40,000 years ago.

Relative dating

Dating of rocks or the fossils they contain by assigning them to their place in the sequence of geological events.

RNA

Ribonucleic acid: the nucleic acid that translates the genetic message of the DNA into protein building. Differs from DNA in containing the sugar ribose instead of deoxyribose, and the base uracil instead of thymine.

Scala naturae

Arrangement of living forms in a scale of increasing complexity, but not a realistic view of how nature is organized.

Sedimentary rocks

Rocks composed of compacted particles derived from pre-existing rocks and transported to their place of deposition by wind or water.

Solutrean

Third major phase of cultural development in the Upper Paleolithic in southwestern France and northern Spain, occuring between 22,000 and 18,000 years ago, but not recognized elsewhere in Europe.

Species

The basic unit of animal classification and evolution, comprising the largest population of individuals capable of interbreeding and producing fully fertile and viable offspring.

Stromatolites

Layered structures formed by the trapping of sediments by colonies of blue-green bacteria.

Sympathetic nervous system

See autonomic nervous system.

Taxon

Category (e.g. species, family, order) at any level of the classification hierarchy.

Tetrapod

Four-footed vertebrates, comprising the amphibians, reptiles, birds, and mammals.

Thermoluminescence

Method of obtaining year-dates. The number of trapped electrons in a sample is determined by heating it and measuring the amount of light given off by escaping electrons.

Virus

Fragments of DNA or RNA in a protein coat that are parasitic on living cells.

For Further Reading

I t has been impossible in this volume to do more than introduce the many fields of science that impinge on the biology and evolution of our species. Following is a list of a few books that explore these various areas of knowledge more deeply or from different points of view. Each also contains references to other works.

Andrews, Peter and Chris Stringer. *Human Evolution: An Illustrated Guide.* New York: Cambridge University Press, 1989.

Bahn, Paul and Jean Vertut. *Images of the Ice Age.* New York: Facts On File, 1988.

Cloud, Preston. *Oasis in Space: Earth History from the Beginning.* New York: W. W. Norton & Co., 1988.

Conroy, Glenn. *Primate Evolution.* New York: W. W. Norton & Co., 1990.

Dawkins, Richard. *The Selfish Gene.* New Edition. New York: Oxford University Press, 1989.

Day, Michael. *Guide to Fossil Man.* 4th Edition. Chicago: University of Chicago Press, 1986.

Eldredge, Niles. *Time Frames: The Evolution of Punctuated Equilibria.* New York: Simon & Schuster, 1985. (Reprinted, 1989, by Princeton University Press).

  Life Pulse: Episodes from the Story of The Fossil Record. New York: Facts On File, 1987.

  Fossils: The Evolution and Extinction of Species. New York: Harry N. Abrams, 1991.

  The Miner's Canary. Unraveling the Mysteries of Extinction. New York: Prentice Hall, 1991.

Fleagle, John. *Primate Adaptation and Evolution.* New York: Academic Press, 1988.

Johanson, Donald C., and Maitland A. Edey. *Lucy: The Beginnings of Humankind.* New York: Simon & Schuster, 1981.

Johanson, Donald C., and Jamie Shreeve. *Lucy's Child: The Discovery of a Human Ancestor.* New York: William Morrow & Co., 1989.

Johanson, Donald C., and Kevin O'Farrell. *Journey from the Dawn: Life with the World's First Family.* New York: Villard Books, 1990.

Jones, Steve, R.D. Martin, and David Pilbeam. *The Cambridge Encyclopedia of Human Evolution.* New York: Cambridge University Press, 1992.

Leakey, Richard, and Roger Lewin. *The Making of Mankind.* New York: Dutton, 1992.

  Origins Reconsidered: In Search of What Makes Us Human. New York: Doubleday & Co., 1992.

Lewin, Roger. *Bones of Contention: Controversies in the Search for Human Origins.* New York: Simon & Schuster, 1987.

———. *In the Age of Mankind: A Smithsonian Book of Human Evolution.* Washington, D.C.: Smithsonian Books, 1988.

Margulis, Lynn, and Karen V. Schwartz. *Five Kingdoms: An Illustrated Guide.* New York: W. H. Freeman & Co., 1988.

Pfeiffer, John. *The Creative Explosion.* New York: Harper & Row, 1982.

Poole, Robert M., Editor. *The Incredible Machine.* Washington, D.C.: National Geographic Society, 1986.

Savage, R.J.G., and M.R. Long. *Mammal Evolution: An Illustrated Guide.* New York: Facts On File, 1986.

Tattersall, Ian, Eric Delson, and John Van Couvering. *Encyclopedia of Human Evolution and Prehistory.* New York: Garland Publishing, 1988.

Watson, James D. *The Double Helix.* New York: Atheneum/Macmillan Co., 1980.

White, Randall. *Dark Caves, Bright Visions: Life in Ice Age Europe.* New York: American Museum of Natural History, 1986.

Willis, Delta. *The Hominid Gang: Behind the Scenes in the Search for Human Origins.* New York: Viking, 1989.

Index

Italics page numbers indicate illustrations